Satellite Spotting for Youngsters

Peter Bassett F.R.A.S.

An inter-active book via our website.

Astronomy Roadshow Publishing
Further paperback copies, signed copies e-books with direct internet links can be purchased from…

www.outerspacebooks.com

Our planetarium service
www.astronomyroadshow.com

Our YouTube Channel   @astroroadshow

All of the links in this book plus many more are offered as a short cut via the satellite-spotting website. Just search the relevant page.

Updated May 2025

All rights reserved. No part of this book may be reproduced by any means, graphic, electronic, or mechanical without prior permission of the publisher except embodied quotations.

Credits to the images have been given wherever possible; the author took around half. This book has been in the making for forty years. I thought it was about time to release it.

	Contents	**Page**
1	Introduction	4
2	In the Beginning	6
3	A Free Hobby, change the world	13
4	If you want to spend	17
5	How the Professional do it	21
6	Comfort Observing	25
7	Why can we see Satellites?	28
8	Solar power for Satellites	35
9	Satellite Orientation	39
10	What is an Orbit?	43
11	Rocket Launchers	48
12	Launch Sites	50
13	Types of Earth Orbit	54
14	Getting into Space	60
15	Ground Track	65
16	Satellite Types	68
17	Space Weather	78
18	Space Junk	82
19	How Bright?	89
20	Lighting Problems	94
21	Visual Sighting Phenomena	98
22	Observing Techniques	108
23	Which satellite is which?	110
24	Satellite Track Photography	119
25	Moving Image Recording	130
26	Telescopic Imaging	138
27	Simple Image Processing	142
28	Favourite Targets	147
29	Transiting Satellites	164
30	Re-entry	168
31	War in Space	179
32	The Future in Orbit	185
33	Stamps and Covers	188
34	After-thoughts on Satellites	190
35	Further reading / links	194
36	Other Books by the Author	196

Most Satellite projects have a designated patch that represents the mission. Once a particular satellite is witnessed, purchase the corresponding patch and wear it with pride. Such patches may be obtained on eBay.

Chapter 1 Introduction

Satellite Spotting is a space age version of Train or Plane Spotting. It is possible to watch satellites passing over your hometown night after night. They look like moving stars of varying brightness. Dozens of satellites can be seen every clear night passing over silently. We may be completely in the dark down here, but hundreds of miles up, the Sun still shining on satellites. Anything larger than a car that is less than 600 miles up can be viewed without any binoculars or telescopes. This hobby can be completely free.

My first realisation that satellites can be viewed at all with the naked eye was back in 1975. I kept a daily diary for several years in order to keep a record of little achievements for when I grew up (I am still waiting).

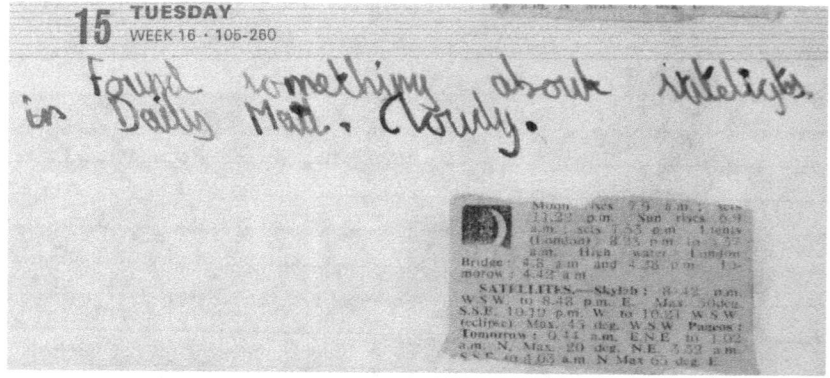

This book is written for the complete beginner; perhaps a youngster at a school with no money looking for a hobby or career. Perhaps you just want to sit on a deck chair at night and watch these things go silently by as hedgehogs snuffle on the ground (I have an e-book /

paperback on Amazon about those too – Attracting and Looking After Hedgehogs).

A Safe Occupation...

Plane spotting can get you into trouble. Several people have been arrested in various countries accused of spying. However, all they were doing was jotting down plane types and numbers just as a record that they have seen them. Some people do the same for trains or even 'Eddie Stobart' Lorries; each one has a woman's name printed on the cabin. Satellite spotting will not get you into any kind of trouble... well there is actually an exception; this is covered later.

The author with his dedicated satellite recording cameras. These are not essential pieces of equipment; your eyes will do nicely

This book is aimed at youngsters from around 10 years old up to GCSE students.

**Please refer to www.outerspacebooks.com
For clips & updates. All links in this book is available via the website. Use the pages connected with the 'Satellite Spotting' menu.**

Chapter 2 In the Beginning…

To reach a height beyond the atmosphere and to have any hope of achieving a minimum orbit, (apart from one or two exceptions) only liquid fuelled rockets can supply that immense power. The world's first rockets of this kind were developed by Robert H. Goddard in the 1920's & 30's. A complete prototype of such a rocket is on display at the Goddard Museum of Roswell, New Mexico USA… the same place as the suspected Flying Saucer crash of 1947.

Photo by P. Bassett taken in Roswell, New Mexico.

The space age truly began in 1942 when the first rocket reached the edge of the atmosphere. It was called the A3, the chief designer was the German rocket engineer Wernher von Braun. This liquid fuelled rocket reached 80km; the very edge of space itself.

The A4 was invented soon after and reached 100 km and was renamed as the V2 rocket by Dr Goebbels. None of these was powerful enough to enter orbit, but did succeed reaching space. Around 3000 were fired toward England in 1944 - 1945; the main targets were London, Norwich & Ipswich. There was no defence against these supersonic vehicles.

Wernher Von Braun once said to his dentist "There is nothing wrong with these rockets, just the target, it should be the Moon." The Gestapo arrested him for this and other comments but released by a letter from Hitler himself as he was so crucial to his war.

After the war, many V2 rockets were captured. British scientists under Operation Backfire launched three from Germany. Later around 200 German rocket scientists moved to the USA along with over 100 rockets. The

Soviet Union also captured as much as they could but just failed to gain the quality of information and scientists that the US had.

V2 rockets may be viewed in museums around the world. This one is at Washington DC Air & Space Museum just a few hundred yards from the Capital Building... Photo by the author.

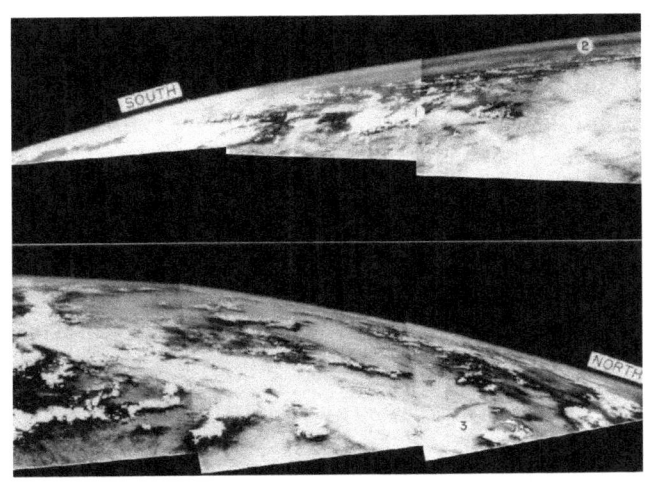

The very first pictures ever taken from space, a V2 rocket was used. 24th October 1946, launched from White Sands, New Mexico, USA.

A clip is included on the website.

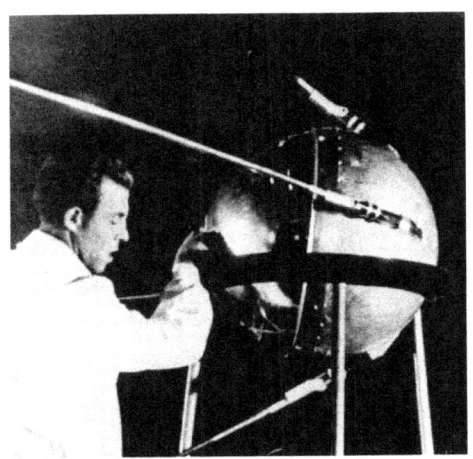

Above; the world's first artificial satellite launched by the Soviet Union in 1957. It remained in orbit for several months. It was a faint object to see but millions witnessed the first artificial satellite as a moving star in the sky.

Most of the V2s reallocated to White Sands Air Force Base in New Mexico and seven to Fort Bliss, Texas. Von Braun assembled his team at Fort Bliss and tried to convince the US Army to adapt it for space missions.

V2 launches became more ambitious with cameras being placed in the nose and other instruments such as temperature gauges & barometers. Rhesus monkeys flew on several launches to prove that an artificial capsule can support a living creature in space.

By August 1953, the first Redstone rocket that had two stages was launched at Cape Canaveral, Florida. It just about reached space and landed several hundred miles away into Atlantic Ocean. As Von Braun's skills at handling politicians increased, he held secret meetings with top Naval officers in Washington DC. His idea was for the Navy to build a small scientific satellite and for his Army Redstone Rocket with an extra upper stage to launch it into orbit.

It was decided that a new rocket built by the Army Missile Agency would be provided instead; a largely untested Vanguard rocket, would be employed rather than Von Braun's reliable Redstone. It did not make sense but hey Ho!

In the meantime, the Soviet Union (Russian group of countries) shocked the world by launching Sputnik 1 on the 4 October 1957; the space age had truly begun. Observers saw the very first artificial satellite passing over as a faint moving star. It remained in orbit for four months as they prepared their next mission. On 3 November 1957, they launched the first dog into space, Laika.

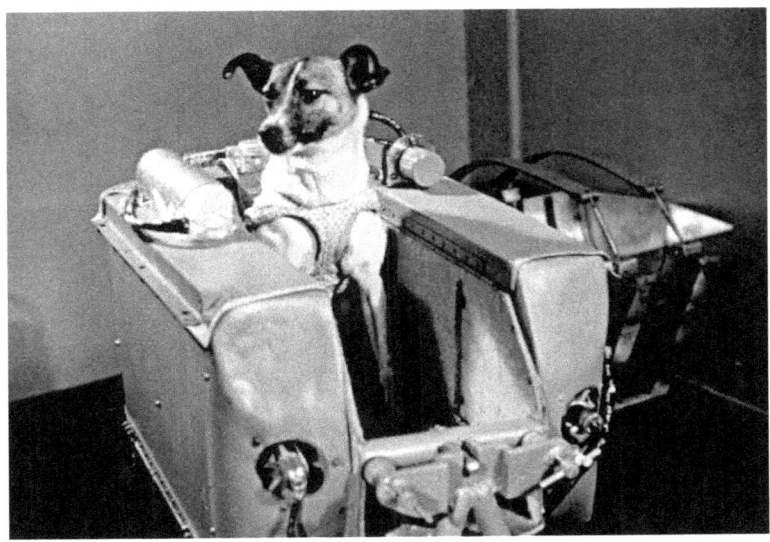

The American Vanguard rocket was on the pad at Cape Canaveral and on 6 December 1957, it was launched. Millions of viewers held their breath as the rocket rose 6ft in the air and just blew up.

Von Braun was asked what he needed to get an American satellite in orbit in 90 days. He replied *'A Redstone rocket and I can do it 70 days.'*

The upgraded Redstone rocket was a converted military missile, so to give it a civilian feel it was re-named Jupiter C. After 84 days, on 31 January 1958 the Explorer 1 satellite was finally placed into orbit. The USA had now joined the space age.

The ladies team above was responsible for the tracking and orbit calculations

Chapter 3 A Free Hobby that can change the World

Anyone at school even without any pocket money can take up this space age interest. Show off to friends that you have seen a $1 billion satellite called Envisat or even better the International Space Station pass over your garden the previous night and waved to the astronauts. This would not be some wild story, but a true reality.

In the UK alone, some 70,000 people are employed designing, building and operating satellites. An extra 100,000 related jobs will be created in the next few years. Perhaps some youngsters at school today may gain a spark of enthusiasm in from satellite spotting and consider something related to it as a career. Britain produces some of the most reliable satellites in the world.

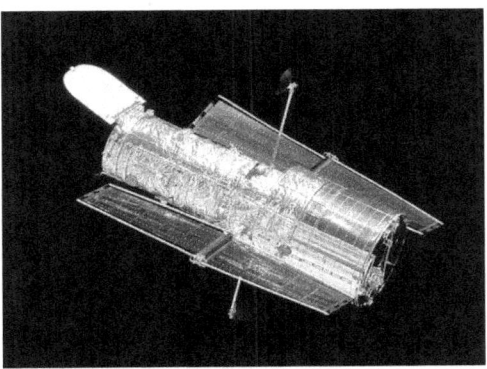

British Aerospace (now Airbus – Defence & Space) built the original Solar Panels for the Hubble Space Telescope in Bristol. The camera sensors built by Teledyne Technologies also in the UK.

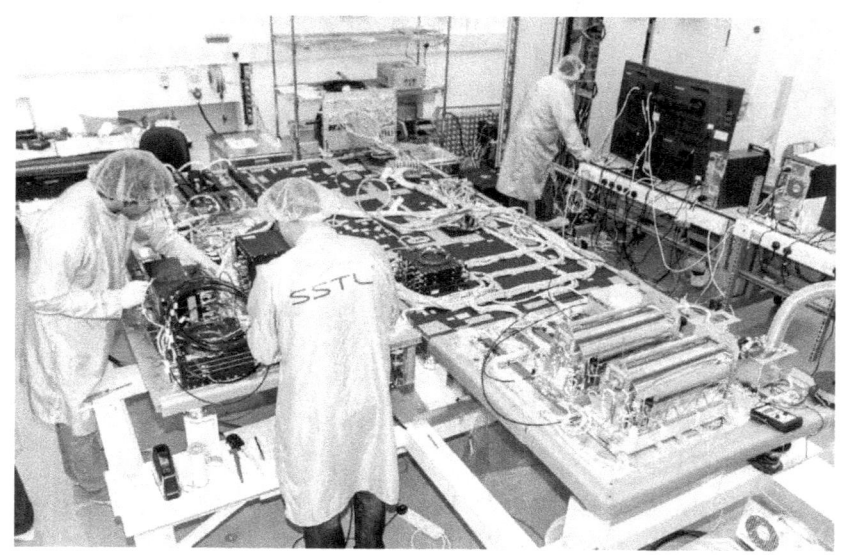

Technicians working on a satellite; Surrey Satellite Technologies LTD, Guildford, Surrey, UK.

Working in the space industry has helped a company called Tessella in Oxfordshire to recruit some of the best brains in the business. One of the owners, Mr Whittle, once said…

"From when we are kids, it's all about space and dinosaurs. Would you rather work in a bank or somewhere you can make a real contribution to your planet?" (His words not mine; I do not want my bank account frozen).

A career in the space industry (or banking) can involve making good money, but producing a difference to the world and change future history in a good way is priceless. Work is not just about making money, making an impact and changing the world for the better is far more important.

If you do not happen to live in the USA, please do not feel that it will hold you back. The USA does indeed produce some amazing technology and is well known as a space aiming country. Remember the following simple points;

Sir Isaac Newton (British) discovered the laws of motion that control orbits.

The original idea to get into orbit; Konstantin Tsiolkovsky - Russian; in case you did not guess already.

The first person to write about communication satellites; Arthur C. Clarke - British.

The first rockets into space; Wernher Von Braun - German.

The very first satellite - Russian.

The practical method that was chosen to reach the moon; originally formed by the British Interplanetary Society, not NASA.

The most distant landing to ever take place, Saturn's moon Titan - European Space Agency, that includes France, Germany, Belgium, UK etc.

The only landing on a Comet so far - European Space Agency.

This could go on page after page. Careers in satellite and spacecraft design, construction, launch, communications, data handling, tracking etc are worldwide. Well over five million jobs are directly

15

related to this exciting field that is changing the way we live and look at our fragile world.

The Space Shuttle Enterprise landed at Stansted Airport, Essex, UK in June 1983.

It is sometimes hard to relate to distant galaxy millions of light years away, but seeing a real spacecraft this close or even shaking the hand of an astronaut is very different. The Enterprise shown here never actually flew into space, but was good enough for us – my friends and I became known as the 'metal-heads.' Author on the right.

Chapter 4   If you want to spend…

As mentioned in the previous chapter, Satellite Spotting can be a free hobby. However, if you really must spend here is a simple guide that can take your interest to a higher level. I will always encourage spending as little as possible though.

Binoculars
If you do not own binoculars, purchase a used pair. eBay is a perfect place; ask someone who is registered to get a pair for you, there are thousands of them. Buy used things; it is the best form of recycling.

The most common problem with used binoculars is the misalignment of the internal prisms due to a hard drop on the ground. No matter how much you mess with the settings, the two images never line up and you will end up with a real headache looking through them. Just ensure that they are aligned and the lens surfaces are in good condition (no scratches) before purchase.

Binocular power is expressed in two ways. Example: 8x30 means the magnification is 8x normal vision; and the size of each lens at the front is 30mm across. The larger lens, the fainter satellites you can see. With magnification; the higher the power, makes it more difficult to find and follow moving objects, a lower power is best. It will also give a brighter image and less shake from your hands.

An ideal pair of binoculars for a beginner is 7x50. Any satellite that is larger than a small wheelie bin (trashcan for USA) may be seen up to around 2,000 km up (1,300 miles. My largest pair owned was 25x125. This allowed me to see satellites as far as around 36,000 km (22,000 miles). The binoculars I now mostly use are 7x50 and 12x70.

I did manage to catch a glimpse of the Space Shuttle Columbia on STS9 mission of November 1983 with my high power 25x125 bins and did see a definite shape to it. John Young was the commander on board; the only astronaut to have walked on the Moon and fly the shuttle. I did wave to him.

25x125 Binoculars by Vixen; powerful enough to see satellites up to 36,000 km. I named them Skylab after my first photographed satellite.

Telescopes

Normally telescopes are useless for most amateur satellite tracking. They move too fast to follow with a telescope. However, there are areas where they become essential.

At the beginning of the space age, Russia, Britain and the US produced specialised cameras built into telescopes to track the orbits of satellites accurately and even attempted to image their shape; some of these details are still secret.

Technology of this type is now in the hands of serious amateurs. Satellites can be tracked accurately with software controlling the drive motors of the telescope. Then it is magnified to reveal the shape of the satellite itself. A recording can be made via a webcam or a specialised telescope camera. This can then be cleaned up for a final picture or video.

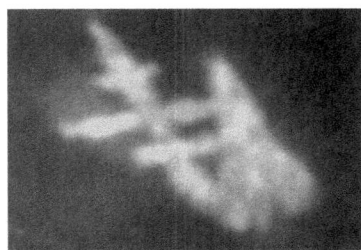

The space shuttle Atlantis docked with the Russian MIR space station in 1995; Photo *taken by a satellite spotter.*

As mentioned in the introduction of this book that Satellite spotting is a safe hobby, but if you take pictures of top-secret spy satellites and then publish them, you could be inviting the police onto your doorstep within days.

Cameras / Requirements
The exact details of how to record satellites will be dealt with later. A camera for satellites should be...

Digital Single Lens Reflex (DSLR). Include a standard lens (or wide angle; a bigger advantage)

Have ISO capability to at least 1,600 (ISO determines how sensitive the Imaging chip is, the higher the number the fainter image it can record more quickly.

At least 16MP resolution. The higher the number, the easier it is to record faint satellites.

Capable of exposure times of up to 30 seconds. (The shutter will remain open exposing to the sky for 30 seconds without touching the camera).

Ability to lengthen the exposure time further still via Bulb setting (B-setting, in other words you set the exposure time manually, several minutes if you want, just keep your finger on the remote trigger button).

Have a standard tripod socket on the bottom.

Capable of either a remote control start of an exposure or cable release to avoid jerking the camera.

It is possible to obtain such a camera for around £100-£250 ($150-$375) if second hand - £250-£500 ($300-$600) new as a rough guide. I have used a Canon 600D; as of 2017, I upgraded to a Canon 1300D.

Chapter 5 How the Professionals Track Satellites!

There are several forms of satellite tracking by the professionals. The reasons for such data vary;

Studying the atmosphere. Very low air pressure still exists in low Earth orbit; this causes a slight drag on the satellite and reduces altitude over time. They are all destined to re-enter the atmosphere.

Activity on the Sun changes our upper atmosphere. These effects alter a satellite's orbit.

Measuring the exact distance of a satellite to confirm Einstein's laws of gravity.

Monitoring the orientation of a satellite if control is partially lost.

Fixing the exact position of Geo-stationary satellites to ensure no collisions take place with others.

Imaging the condition of a spacecraft.

Spying on the design of foreign secret spy satellites for potential violation of weapons treaties. It is currently illegal for any nation to deploy weapon systems in space.

Laser Tracking

Some satellites have reflecting glass reflectors on their surface called corner reflectors to bounce back any light that falls on them from any angle; rather like cat's-eye studs in the road. Next picture shows a spare Lageos satellite on display at Marshall Spaceflight Centre, Huntsville, Alabama. The original was launched in 1976 to an orbital altitude of 6,000 km.

The telescope shown next is a laser-firing instrument to a Lageos in orbit. This one is based at Herstmonceux in Sussex, England. It began working in 1983. This is still in operation as of December 2021. This particular experiment discovered that England was moving toward France at a rate of 15 mm a year. This fact became important in the design of the Channel Tunnel. **LAGEOS** stands for **La**ser **Geo**dynamics **S**atellite.

A clip is included on www.outerspacebooks.com

A laser being fired to a reflector on the moon. Photos by McDonald Observatory and NASA.

Visual Tracking

A dedicated optical telescope / camera system was designed in the 1950s to photograph the track of a satellite and compare its position against the stars.

These telescopes are called Baker-Nunn Cameras, specially designed to track satellites.

Later versions of these telescopes could zoom right in and record structural detail of the satellite too. Details of some of the more modern forms of equipment are still top secret but have been and still used to photograph possible structural problems on spacecraft such as the Space Station.

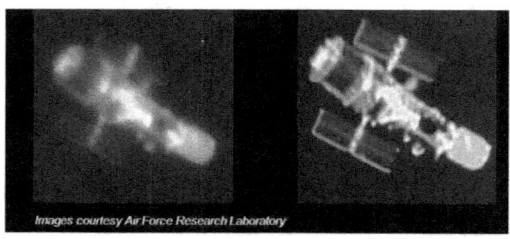

This image is by a US Air Force satellite tracking camera; a clear demonstration of the power of modern satellite tracking telescopes. Such pictures are normally top secret.

If used to its fullest capability, the hole in the wing after the launch of the Space Shuttle Columbia in 2001 could have been detected from the ground. A rescue mission would then have been designed, as the Atlantis was on the launch pad ready for a mission. The crew could have been saved. NASA managers, turned down such an offer of help, some were fired from NASA.

Photo of a Baker Nunn camera with its cover at Alamogordo, New Mexico. Photo by the author.

Chapter 6   Comfort Observing

If time is going to be spent outside watching these amazing machines pass over your hometown, then you need to make yourself comfortable. Coldness can soon set in to your body (especially in the UK) it may also be difficult to read your satellite listing and you will not have an enjoyable experience. You will be wishing you were inside watching a good movie instead and your new interest could be history.

You may just wish to watch a single satellite pass just so you can amaze your friends and say, you have seen Cosmos 2103 with its rocket case trailing and tumbling behind. In this case, just get your shoes / slippers, coat on five minutes before the predicted pass, and pop outside. Get used to the dark, if you see stars, you should see the satellite as predicted. This requires very little planning, you can pause your movie in the meantime and see the space age event take place in the real world rather than seeing a report on the internet.

If however you want to witness dozens pass over on a beautifully clear moonless night, then a little planning will make it a much more enjoyable and memorable experience.

Clothes
In the UK or similar climate, wear a thick coat with either a hood or a separate hat. Most of your body heat is lost through your head. Thick thermal socks and boots keep your feet warm. Do not stand on concrete if possible. If there is no other option, try standing on an old piece of wood. Wear thermal gloves unless you are

setting up your camera; this can be a delicate operation. Do not forget glasses if you need them for adjusting your camera and reading from your satellite list.

Deck chair

If you are looking for faint satellites in high orbit that require binoculars, do consider a deck chair. This will allow long duration observing sessions in comfort, a steady position will allow accurate sweeping and observing for satellites.

Warm coat, hood, gloves, thick boots or shoes and off the ground if possible. This combination produces relaxing and warm conditions for satellite spotting. Position yourself in a sheltered area; away from winds. The binoculars shown are 12x70. Image of the author; ask for a signed photo of this if you desire. At least it shows my best side.

Most of the time I chose to stand especially if I am only out to see one or two targets or if I see clouds approaching fast. It really is not worth the effort otherwise.

Print out & Red Torch (flashlight)

Use www.heavens-above.com for the satellite predictions. A dedicated chapter in this book describes how to use it. Print out the satellite list for that night and highlight the objects you most wish to see. If you want to be flash, try out the 'app' that will give live information on each satellite that can be seen from your location.

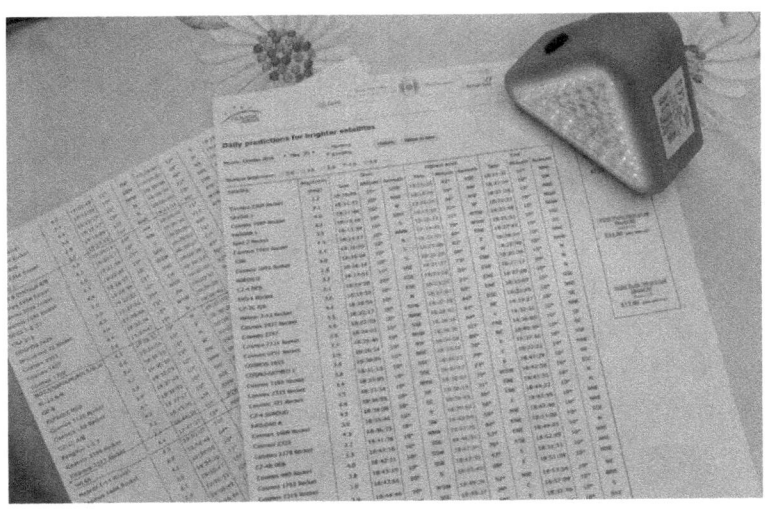

Ensure you have a red torch to light the list. A white light will blinding at night but can be preferred for reading small print if you want. An LED torch will last and last without needing new batteries.

Tick off the satellites after observing them as a record of your achievement. At certain times of the year, over a hundred satellites can be seen in just 4 hrs or so. My personal single evening record is 119 recognised satellites including seeing seven passing over at once.

Chapter 7 Why can we see Satellites?

We can see stars at night as they all shine by the light of their own. The moon, the planets and satellites do not generate light but are lit by the Sun instead. During the night on Earth, we are standing within the Earth's own shadow and the Sun is below the horizon. Stand on a ladder and you won't expect to see the Sun, walk to the top of a hill, still no Sun, fly up in a plane to 4000 metres; nope! Fly up in a rocket to 300km or so and aha! We see the Sun above the horizon again. It is now illuminating you but the ground down below is in darkness with city streetlights, fires and flares from oilrigs.

At sunset, there is a much better chance of experiencing this on a small scale if you feel athletic enough. At the back of my old school (Upbury Manor, Gillingham, Kent UK), there is a 40° inclined hill about 60 metres high that lead down to the town of Chatham. In November and February, the Sun sets around 4pm. I used to walk to the bottom of that hill after school on a very clear day and watch the sunset. Then I ran half way up as fast as I could, turned and watched the sunset again, then ran to the top and saw a third sunset. The bottom of the hill was getting dark while the Sun lighted the top.

Now imagine a ladder, climb up it quick and witness a fourth sunset, then onto a small plane up at 1000 metres, then a jet at 15,000 metres and a satellite at 400 km. Anybody can witness the Sun set many times on the same day if you keep going higher at a faster rate than the setting Sun. The Sky TV satellite is only in darkness for 45 minutes or so per 24hr orbit. Far enough from the Earth and it becomes visible 24hrs a day, but down below that part of the Earth may be in darkness. The deep space probes experience the Sun constantly.

Above; as the Sun has just set to the West, the Earth's own shadow can be observed to the East. The low dark line is the beginning of the shadow within the atmosphere. As the Sun gets lower, this line rises until we are inside it; the night. The full Moon is shown here too. The opposite takes place at sunrise.

The diagram below shows what part of a satellite's orbit would be visible from the ground.

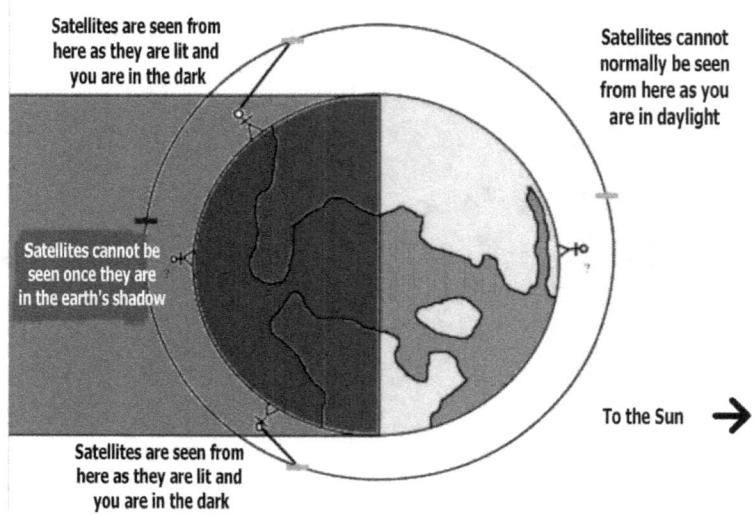

More on the Earth's Shadow

As the Sun is 109 times wider than the earth, the earth's shadow is therefore tapered as a cone. It reaches out to a point 1.36 million km from the earth. Beyond that, the Sun is fully visible apart from a small silhouette of the Earth against the Sun's image. The shadow is in two parts; The Umbra and the Penumbra. The Penumbra is just partial show, only the Umbra is full shadow. Both parts are shown as the Moon passes through the shadow sometimes when it is full. This how is a lunar eclipse works.

The next diagram shows the view of the Earth from space within the earth's own shadow; with the Sun aligned in the background. At point A, the Sun is totally eclipsed. Point B will show the outer solar atmosphere, the Corona, but not the Sun's surface itself; just as in a total solar eclipse from earth. Point C and D are at beyond 1.36 million km from earth. Much of the solar disc is now visible; the Earth will just be a silhouette against the Sun's face. The same can occur on the Earth with Mercury & Venus passing in front of the solar disc, known as a Transit. During such rare occasions, the Earth is actually passing through the Penumbra part of those planets shadows.

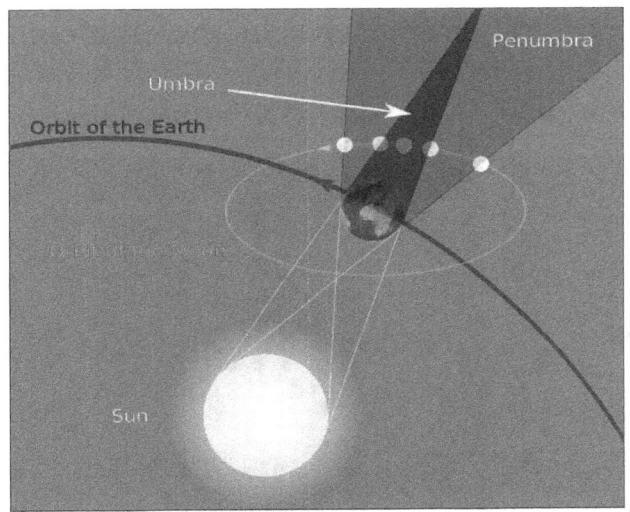

The Umbra of the Earth's shadow is the deep darkening shown on the moon; the surrounding hazy area is part of the Penumbra.

As a satellite travels into the earth's shadow, it fades for a few seconds as it enters the Penumbra area then vanishes completely as it passes into the Umbra. The time this takes depends upon the angle of approach. The satellite may head straight to the shadow at 90° and can fade in just 5 seconds. If it approaches from a shallow angle, it could take 30 seconds or more.

A total lunar eclipse will show the Moon glowing in red light. This is due to the red part of the Sun's spectrum being refracted through the earth's atmosphere around the edge and redirected into the Earth's shadow. The light bends slightly inward and by the time it reaches around 300,000 km, the whole of the earth's shadow is filled with this red light. Since the Moon is 400,000 km away, it bathes in this soft light during eclipses.

September 2015 Lunar Eclipse from the UK; the Moon in almost totally immersed in the Earth's Umbra part of its shadow. The left side is showing the Penumbra.

Seasonal Changes

As the Earth orbits the Sun, its 23° tilt remains pointed in the same direction; the North axis will always point toward the North Star. However, our tilt related to the Sun will change slowly day by day. This effect causes seasons and alters the angle of our own shadow in the sky.

As the Earth rotates, the moon's orbit is tilted in comparison with the Sun, the earth's tilt alters slightly each day compared to the Sun too – seasonal shift; all these changes alter the Moon rise / set position daily as well as its orientation.

During local winter, the shadow is high above us within 2 hours or so after sunset and 2 hrs before sunrise. The summer effect will keep the shadow very low even in the middle of the night therefore satellites would be seen all night with complete crossings. The next picture is of a satellite passing over the UK in June at a few minutes to midnight.

The angle of the shadow compared to your position on Earth defines the number of satellites that can be seen each night. No matter where you are or the season, many satellites will be observed. The number of them and whether you see them enter the shadow or not are defined by those two points; your Position and Date.

This is a satellite called SkyMed 2 passing into the earth's shadow. These fading effects are fascinating to observe. You do not need to be in the middle of Arizona to photograph satellites; although it helps. SkyMed can produce very bright long lasting flares as this did. Normally this satellite is around Mag 3.5 but the Sun reflected perfectly off the flat antenna so it increased in brightness to Mag 1. Photo by the author in October 2014 from his own back garden.

Chapter 8   Solar Power for Satellites

A battery powered the first satellites such as Sputnik 1. As soon as it was launched all the on-board systems were powered from it and drained away its stored electricity. Once exhausted, the satellite is dead; it is still orbiting but no longer functioning. All that design, expense and effort, is now drifting uselessly in an orbit that will eventually decay and burn up. This was extremely wasteful so a solution had to be found.

In 1947, the very first solar cell was invented at the Bell Laboratories in New Jersey, USA. When exposed to light, this thin silicon wafer of various layers would produce a weak but steady current. This power can be used to charge a battery.

A modern Li-ion rechargeable battery suitable for satellites

With the venture into space, new batteries were designed and solar cells / arrays began to become more practical. By the 1970's these became available for domestic use in calculators and charging AA batteries etc. Most satellites use solar power systems, they are now cheap, lightweight and very reliable. Satellites can now be designed to last for many years in orbit instead of a few weeks.

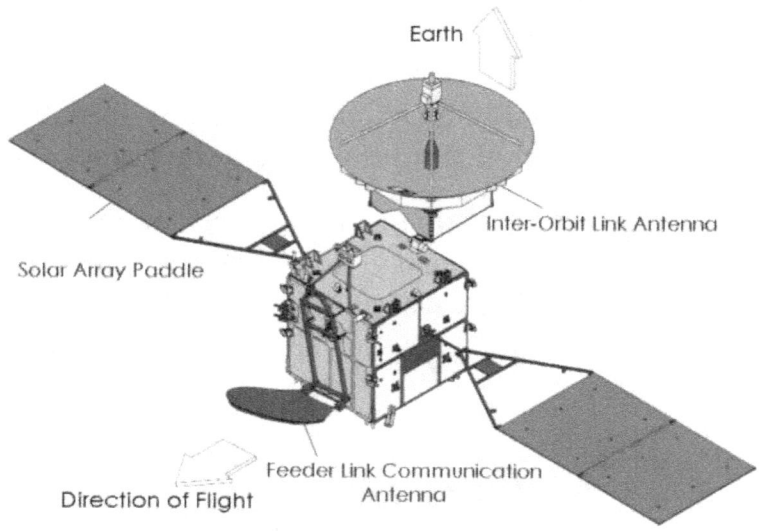

During launch, some systems on the satellite are already switched on and are powered by batteries. Once released from its casing in orbit, the panels unfold and are exposed to the Sun. The batteries are charged for the first time in orbit and the satellite continues to be powered from them. While In the earth's shadow battery power is used only but gets a recharge on the sunlit section of the orbit. This will continue throughout the lifetime of the satellite for up to 10 years or so. The current record as far as I am aware is the British Prospero satellite launched in 1971; still transmitting today. Older satellites are in orbit but not operating.

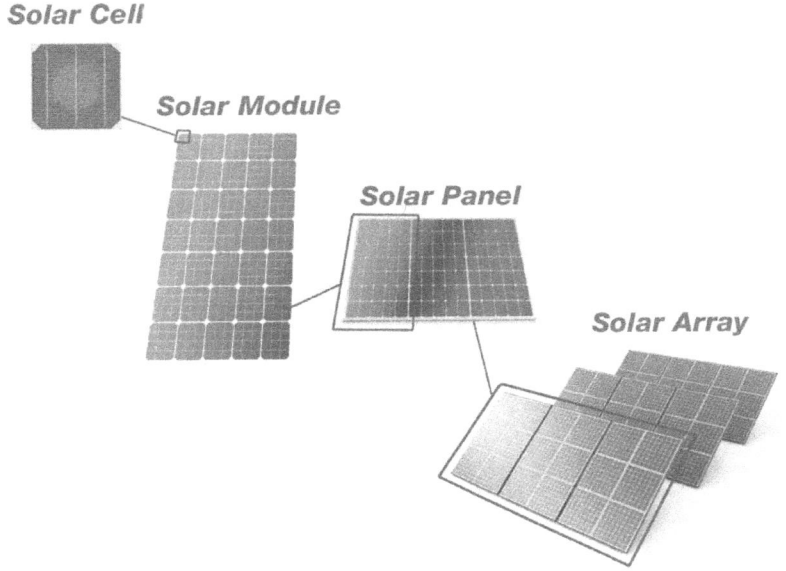

Such array of solar panels can now produce a very useful power supply that can operate in homes, boats and now aircraft. For maximum efficiency, the panels should rotate to follow the Sun; the first satellites to do this were the US Nimbus Earth observation satellites – launched 1964-1978.

Satellites can have huge arrays of cells that generate several kilowatts of power during the daylight part of its orbit. Nothing off course will be generated in the earth's shadow; that is when the battery power is used only. To avoid switching 'glitches' sometimes a satellite is on permanent battery power and is just re-charged when not in the earth's shadow.

A mock-up of a solar array that can be unrolled from a tube or a stack once in orbit. The larger the surface area of such solar panels, the brighter the satellite will be seen from the ground. Images by the author at Cleveland Science Museum, Ohio.

Chapter 9   Satellite Orientation

Every object above the ground such as a plane or spacecraft has several points of reference regarding orientation. Various forces such as air pressure can affect this; wind shear, off axis thrust from propellers, jets or rocket engines. Such changes from a desire path or orientation must be corrected. The following diagram shows the crucial reference points on a rocket but the same applies to an aircraft or satellite in space.

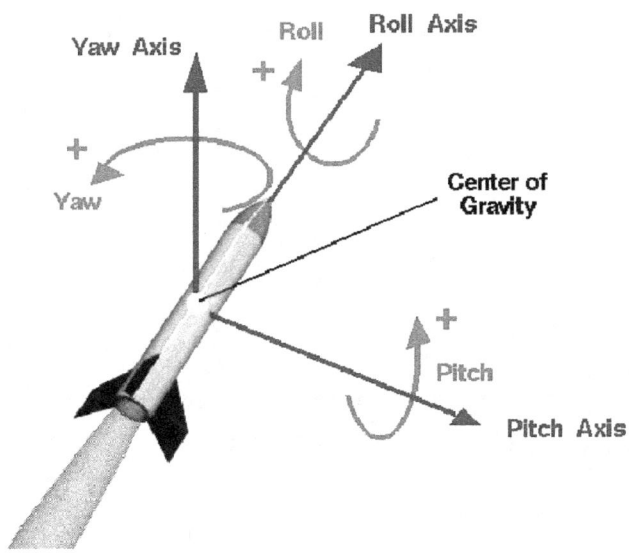

NASA image

Some satellites never require a preferred orientation. LAGEOS for example is just a sphere covered in corner reflectors. A laser from the ground is fired toward it and the satellite just reflects it back. It doesn't matter which way round the satellite is. Others may require precision pointing; for solar power, for observing and for uplink and downlink data.

So how can a free-floating satellite in the depths of space keep pointing in a particular direction?

Thrusters
Small rocket thrusters are built around the satellite, which will allow it to rotate or de-rotate along all three axes. This method is fine if very few such adjustments are expected or if the mission is relatively short lived such as the Apollo Moon missions. Only a certain amount of fuel can be stored so this will define the useful life of the machine.

Spin Stabilising
Any wheel that rotates around its axis will have a natural resistance to change. This is the principal of a gyroscope. If an entire spacecraft were to rotate around itself, then a stable position can be kept without using fuel or any complex technology. The only disadvantage is that it becomes rather tricky for photography and communications if the whole satellite is spinning.

Magnetic Torque
The Earth has a natural magnetic field. The magnetised needle on a compass has a week resistance to change direction. If a magnet is attached on a long arm on a satellite, then the same orientation can take place. Full control can work if three booms are used at right angles to each other with electro magnets on the end. A changing power supply will alter the satellite's orientation compared to the earth's magnetic field. This is ideal for some spacecraft where the operators wish to

keep one side of the machine facing Earth or to a set point in space. This method only requires electricity from solar panels; no moving parts or fuel. The Hubble Space Telescope uses this method with three internal magnetic booms.

Reaction Wheels

Reaction wheels (RWs or sometimes referred to as Momentum Wheels or Inertia Wheels) are fitted inside almost all satellites. It is a type of flywheel, looks like a gyroscope and has similar properties. The heavy flywheel rotates at high speed via an electric motor.

NASA Image.

The Ball Brothers Research Corporation that produced many inventions for World War 2 invented the very first such system. A test in 1955 worked first time to an altitude of around 200 miles. These systems were originally known as Servomotors; the more sophisticated versions today are Reaction-Wheels.

The Hubble Space Telescope is manoeuvred via a combination of reaction wheels and Magnetic Torque. Astronauts on the shuttle replaced the wheels during each service. Imagine engines constantly firing, polluting the space around itself whilst trying to photograph distant galaxies.

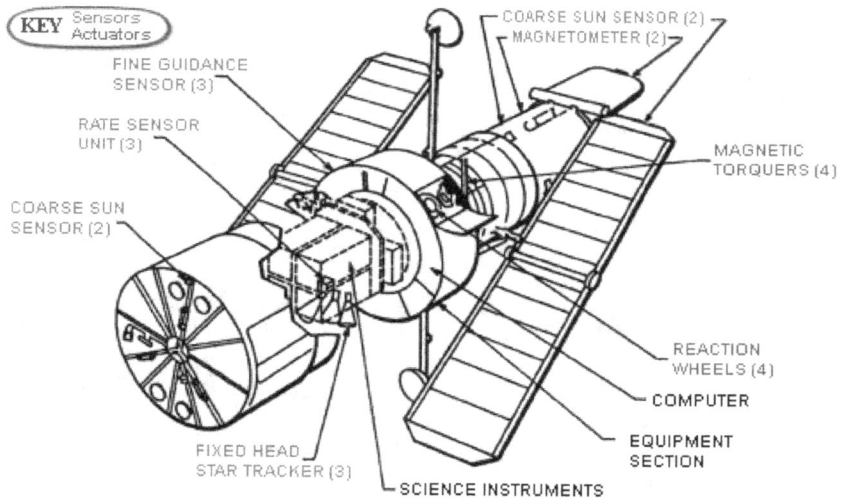

The Pointing Control System (PCS) aligns Hubble so that the telescope points to and remains locked on a target.

Chapter 10 What is an Orbit?

After teaching astronomy full time since 1994, I wish I could have a pound for every time I hear *'Astronauts must have a great time in zero gravity'* or something to that effect. I usually reply *'I am sure it must be, but then if there is no gravity up there then why do they orbit the Earth and not fly away?'*

The reply is usually *'Because of the earth's gravity silly!'* That part is correct; the first is not. So what is going wrong? The astronauts are to blame for this. They are always mentioning zero or micro gravity. Both terms are false. Gravity does indeed get weaker with distance as first proved by Robert Hooke from the Isle of Wight in 1686. He demonstrated this to Sir Isaac Newton and later included it in his famous book on gravity – The Principia... but Newton never gave Robert the credit; it was a bad habit of his.

Therefore, the next question will naturally be *'Why do things float around in space then?'* This is down to motion; satellites are not stationary. In a roller coaster, during the fast descent phase, you may feel that you weigh virtually nothing and you begin to feel you could float out it you were not strapped in. This is a real effect. The vehicle you are in is rapidly descending and so are you. Gravity is doing its job just the same and yet you feel lighter. If you could weigh yourself at this point, you really would have been on an instant diet; kilograms have vanished. Once you reach the bottom of the ride, full weight is resumed; sorry, (unless you have thrown up).

The astronauts are experiencing the same effect at 100% efficiency. Braking systems, rails, or even the air is not slowing them. They are *freefalling* around the globe in a never-ending loop instead of a small dive lasting seconds on your roller coaster. The gravity is still acting upon the astronauts but in a freefall state, you will not feel it. Other examples are fast descending lifts, or best of all; the 'Zero G' plane in the USA... aahhrrrrr back to that again! It is a perfect name for marketing and the company directors do realise that they cannot switch off gravity.

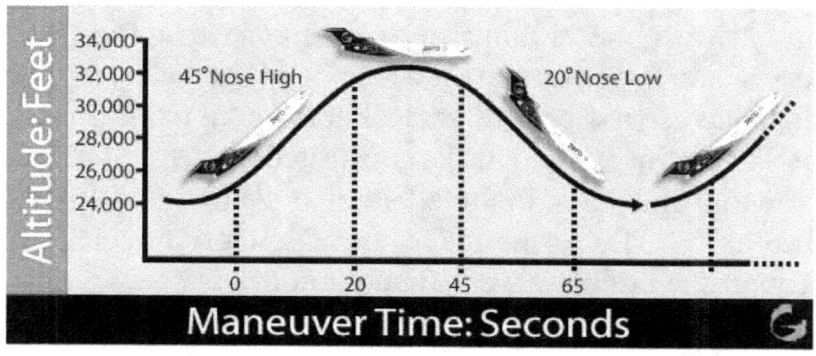

As the plane dives downward everyone on board experiences weightlessness due to freefall / zero weight, not zero gravity.

If you want to experience this for real, the flights in the USA are around $4,500. A similar service is being planned for Europe at around £2,000.

Achieving an Orbit

An orbital path is simply a much bigger example than the Zero G plane mentioned. This is simplified due to the fact there is no air resistance above the atmosphere to slow things down and complicate matters.

If the Earth had no atmosphere, then all you need to do is place a massive canon on top of a mountain and fire a ball horizontally. It may travel 5 km but will begin to drop down as soon as it leaves the canon. Increase the power and the ball will travel further before hitting the ground. Fire it fast enough and it will travel all the way round the Earth and hit the back of the canon. Better still, take the canon down after firing and the ball will return and pass the starting point and go round again forever.

To leave the Earth completely, a special minimum speed is required - 11 km per second. The astronauts that left for the Moon achieved that and any spacecraft heading for the planets etc. require it too.

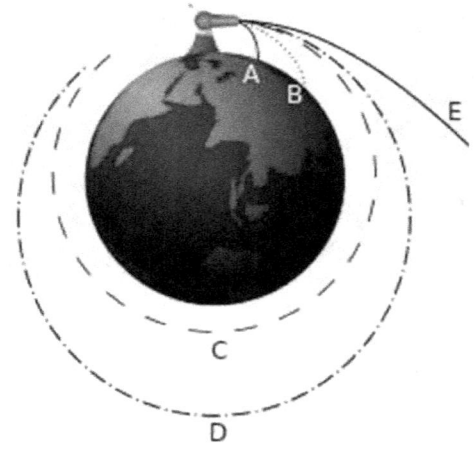

**A=500 metres / sec
B=1000 metres / sec
C=7,8000 metres / sec
D=9,000 metres / sec
E=11,900 metres / sec**

Paths A & B are partial orbits; C & D are complete orbits. E has achieved escape velocity. The speed / velocity is the crucial point here. The gravity of the Earth is always pulling on the orbiting object.

For animations regarding this point, use the link 'Newton's Cannon ball' on the links page of www.outerspacebooks.com

Obviously if you are going to put a satellite into orbit, a canon would be useless. The acceleration would smash all the electronics and kill any astronaut stupid enough to volunteer. So a much gentler acceleration is needed. How can we do that? Well a catapult would be gentle enough, but would not have the energy to push it up to 7,800 metres /second. The air would also slow down its progress until it runs out of energy. The only answer is a rocket!

Orbital times

The higher the orbit, the longer it takes to complete it. This is for two reasons;

1) Because the further up a satellite it, the further it has to travel around the globe.

2) The further the satellite is away from the earth, the weaker the gravity, so less energy is required to keep it up. Travel faster than required for a certain height and the satellite will change to a higher orbit or even leave the Earth altogether. This is how some satellites alter course. Spy satellites can change orbit often so others loose track of their path.

The minimum orbital period is 88 minutes at an altitude of around 160 km. The International Space Station takes around 100 minutes at 400 km. Geo-stationary orbit takes 24hrs at 36,000 km up.

We accept daily weather forecasts but often forget where this data originates.

Chapter 11 Rocket Launchers

Hundreds of designs of rockets have launched satellites since 4th October 1957. Their general design has remained the same but has been upgraded and perfected. This idea keeps costs and failures down to a minimum. The launchers are built in a factory by bulk order; dozens at a time.

The USA on the other hand constantly strives for new technologies. Competition between the Army, Navy and Air Force as well as companies has produced a wide variety of rockets. This policy does indeed produce new technologies but will require several test launches for every new design to prove reliability. Failures are more common until all the problems are sorted out.

A Soyuz Launcher – R7; These are still used for crews and satellite missions.

The top section may change but the booster has become standardised and perfected.

Launchers today are built and used by many nations. The UK has had a success in 1971 with the Black Arrow rocket. It launched a satellite built in Stevenage that is still operating. The factory was in Cowes and test stands are at Alum Bay on the Isle of Wight.

Once a new launcher has been proved successful with a positive record of accomplishment, then several versions of the same rocket often follow. China for instance has produced a range called Long March Rockets.

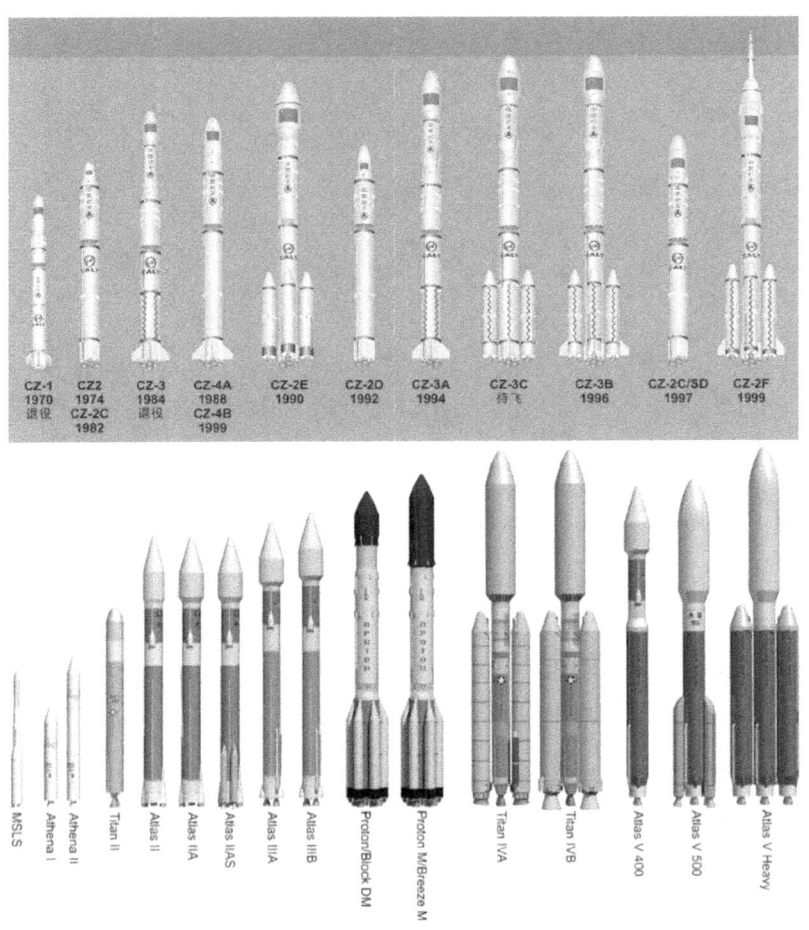

Current US Lockheed launcher range

Chapter 12   Launch Sites

The very first launch site in the world that threw something into space was at Peenemünde; an island off the Germany Baltic coast. The V2 rocket was launched on 6 October 1943.

Relevant clips are on the website.

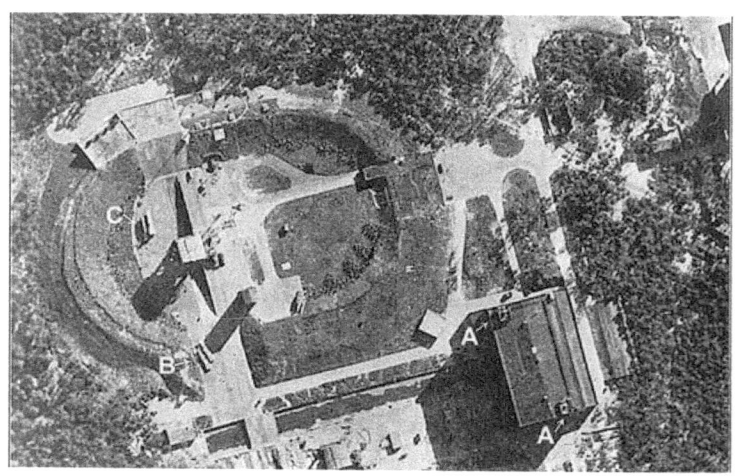

Test Stand V11 at Peenemunde; Image by the Royal Air Force.

The space age truly began in the Soviet Union with the launch of Sputnik 1 on 4 October 1957; the first artificial satellite. This base is now referred to as Baikonur in Kazakhstan.

Kennedy Space Center, Florida

Situated on the mid-Florida east coast, this is the most visited launch site in the world. Launch times and dates are widely publicised on-line and viewing tickets from within the area are available for a small fee. Never underestimate the crowds and allow plenty of time to get past the entry point park up and gain a good spot.

Six hours ahead of any launch is recommended. Other good places include Cape Canaveral Harbor that includes gifts shops and cafes, Cocoa Beach and Air Force Road (my personal favourite).

A satellite launching from Satish Dhawan Space Centre, India

Vandenberg Air Force Base, California
This is strictly a military facility and does not offer tours of any kind unless you happen to be closely related to a member of staff. Open days are arranged for such people.

Launches of spy & cyber related satellites and test vehicles take of from in a secure base many miles away from the public. Due to its location, only polar orbiting satellites can start their missions here, all others take place from Kennedy Space Center, Florida. Boosters falling down in Colorado or Kentucky are bound to attract complaints

US Air Force image

Ocean Odyssey

A Mobile launch platform offers several advantages and launches from them are increasing. The Ocean Odyssey is actually a ship can travel to any part of any ocean in the world.

Four nations share the Ocean Odyssey; Russia, Ukraine, Norway and the USA. The first launch was a DemoSat in 1999. This launch proved the concept would work without launching a real satellite that may have spent years build only to fail.

Air Launch

Micro Satellites can be air launched via a system called Pegasus. A solid propellant rocket is taken up to a high altitude and Pegasus is then released. The solid rocket motor fires for several minutes and places the satellite in low Earth orbit. As of 2014, around 40 launches have taken place using this method.

Above; an early Pegasus launcher photographed by the author in 1992 at the Edwards Air Force Base, California.

NASA image.

Chapter 13   Types of Earth Orbit

The following terms should be learned in order to understand the most common descriptions of a satellite's orbit.

Sub (Partial) Orbit
The V2 rockets in World War 2 achieved the first true partial orbit to the edge of space. At Wallops Island USA, Fairbanks Alaska, Norway and Canada, rockets are still sent on sub orbit hops and back down. These are to test new satellite hardware, research on the Northern Lights, magnetic field, etc. Such vehicles are known as Sounding Rockets. Many amateur rocket builders have achieved Sub orbit and got cameras to reach the edge of space.

Low Earth Orbit (LEO)
The lowest orbit that is practical is 160 km and is completed in just 88 minutes. The maximum altitude described, as LEO is 2000 km; 127-minute circuit. The majority of visible satellites (without the use of binoculars etc) are in this range.

Mid Earth Orbit (MEO)
Only around 15% of satellites are in MEO. This is a range from 2,000 km and below 36,000 km. As they are so high up, they will be faint to observe but not impossible. Binoculars will show many of them. At 22,200 km, an orbit takes 12 hrs, ideal for navigation and communication satellites.

High Earth Orbit (HEO)
All Earth orbits above 36,000 km (Geo Stationery) are considered HEO. These all take more that 24hrs to

circle the earth, so instead of travelling generally West to East, these seem to go East to West. This is only due to the fact we are rotating in a faster time than the satellite.

The VELA 5 satellite in the clean room shortly before placing in its rocket. This is in High Earth Orbit.

Highly Elliptical Orbit

These orbits are generally below 1000 km at minimum altitude and above 36,000 km at maximum. Some communication satellites use this kind of orbit as the majority of its path will appear to remain almost stationary in our sky. Some are achieved by accident if a final firing of a thruster fails and leaves the satellites in an unintended elliptical path around the earth.

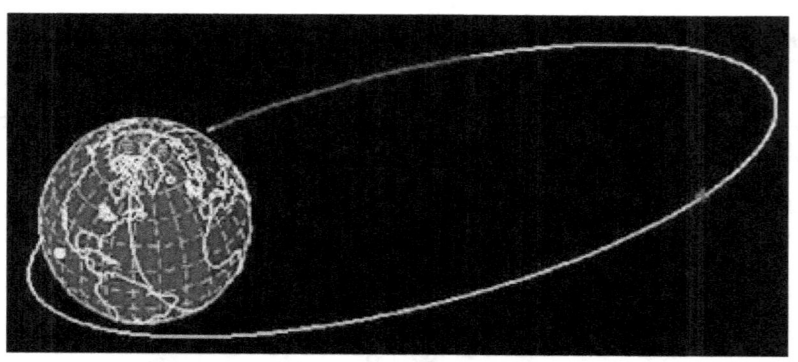

Geo – Stationary Orbit (Clarke Orbit)

The higher the orbit, the longer it takes to complete the circuit of the earth. At 36,000 km, it will take 24hrs for one single orbit. As the Earth also rotates in 24hrs, the satellite, once arrived, will remain above the same point on Earth forever. Such satellites occupy a single ring above the Equator only.

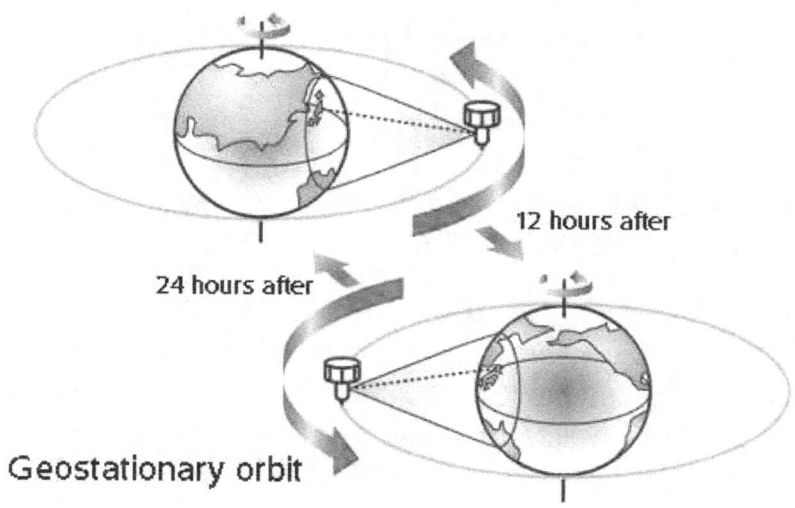

All satellite dishes such as these point to a geo-stationary satellite such as Sky above the equator. Dishes were much larger up to the 1980s as radio wavelengths were mostly used. These frequencies are contaminated by background noise from the stars etc so a stronger signal from a bigger dish was required. If you observe the sky in Microwave, it is almost completely noise free. Most satellites from the 1980's used Microwave transmitters and so smaller dishes were then used.

Microwave receivers are smaller than radio as the signal suffers from less background noise. Lower power transmitters and receivers become possible.

Radio Satellite receivers in South Dakota, USA.

For a single service to cover the whole world, three satellites 120° apart are required. 3 x 120° = 360°.

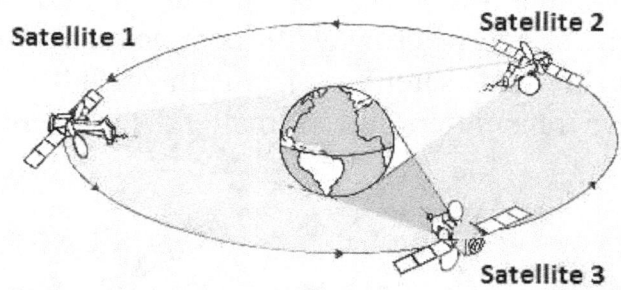

When they run out of thruster fuel, the satellites are at the end of their useful life as they are no longer able to keep in their allocated position.

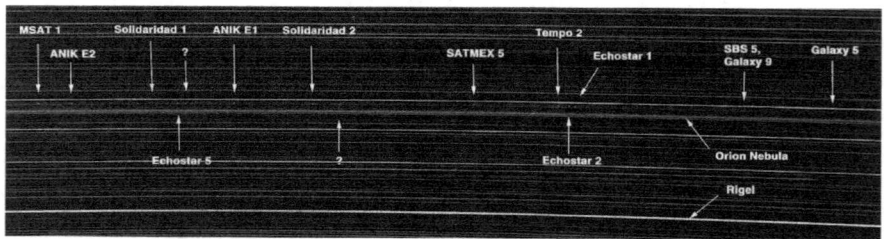

Geo-stationary orbiting satellites; the camera is kept still while the stars trail as the Earth rotates. Picture by Bill Livingstone of Tucson, Arizona.

Equatorial Orbit

An orbit that passes constantly above the equator only. All Geo-stationary satellites are in Equatorial orbits. These are best achieved from a mobile launch platform (Ocean Odyssey) placed directly on the equator to save on fuel and complex orbital manoeuvring.

Inclined Orbit

The inclination of a satellite's orbit is the angle that the orbit crosses the Equator. If a satellite has a 0°

inclination, then it would be orbiting directly over the Equator. If a satellite has a 90° inclination, then its orbit is at right angles to the Equator and it would pass over the poles instead. All other orbits are therefore between 0 °and 90° to the equator.

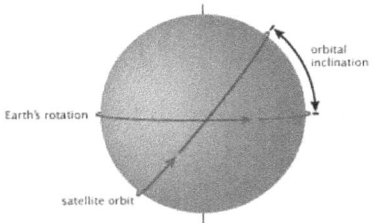

Polar Orbit

The satellite passes over the North and South Poles on each orbit and eventually passes over all points on Earth as the Earth rotates underneath in this orbit. A pass can be seen as North to South, but after a few orbits, the same satellite will be seen as South to North then back again. This is only due to the fact the Earth has rotated so the observer will see the opposite side of the orbit instead.

Sun-synchronous polar orbit

This is a special kind of polar orbit. When travelling in this orbit, a satellite not only travels over the North and South Poles, but it passes over the same part of Earth at roughly the same time each day; essential for some daylight observations.

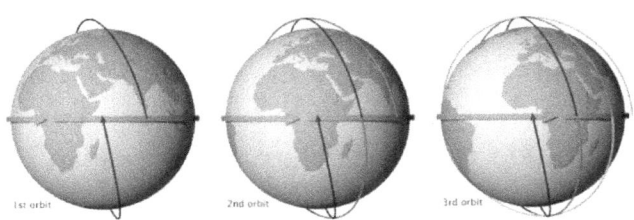

Chapter 14   Getting into Space

The launch from the pad is just the start of a satellite's journey into space. There are many designs of rocket engines today but they all essentially serve the one single purpose; Newton's Third Law of *Motion 'Every action has an equal and opposite re-action.'*

The engine's fuel is mixed with another chemical or gas to improve combustion. These liquids are often super cooled gases to allow extra compression. Most engines circulate the cold fuel around the engine nozzle through pipes skirting the outside then back to the fuel pump. This prevents the nozzle from melting and it pre-heats the liquid into a gas that is more readily combustible. The thrust (measured in kg of force) must be more than the total weight of the rocket itself; it will not go very far otherwise.

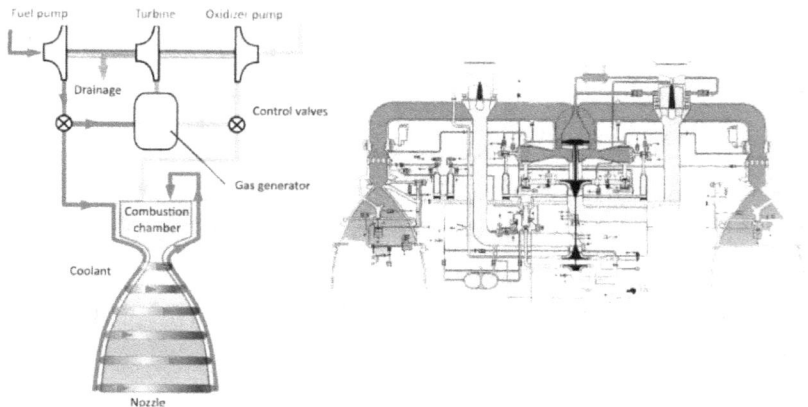

As the fuel is used second by second, the rocket becomes lighter which in turn allows acceleration. The higher altitude that is achieved, the lower the air pressure / air resistance and so allows further acceleration for a second reason.

The engines can gimble off axis to redirect the rocket from vertical flight to an orbital path and take into account wind shear along the way.

The V2 rocket did this by keeping the engine still and moving metal paddles within the exhaust flame to redirect the thrust. The paddles did melt due to the high temperature of the exhaust but survived long enough for short flights.

Four Vanes as the one above painted red directed the flame, which altered the heading of the V2 rocket. The high temperature of the exhaust gradually melted the vanes. As the flights were only designed to last a few minutes to its maximum altitude before falling to its target, this didn't cause a problem.

The longer duration burns for longer partial orbits or full space flight required a new approach; manoeuvre the entire engine nozzle instead.

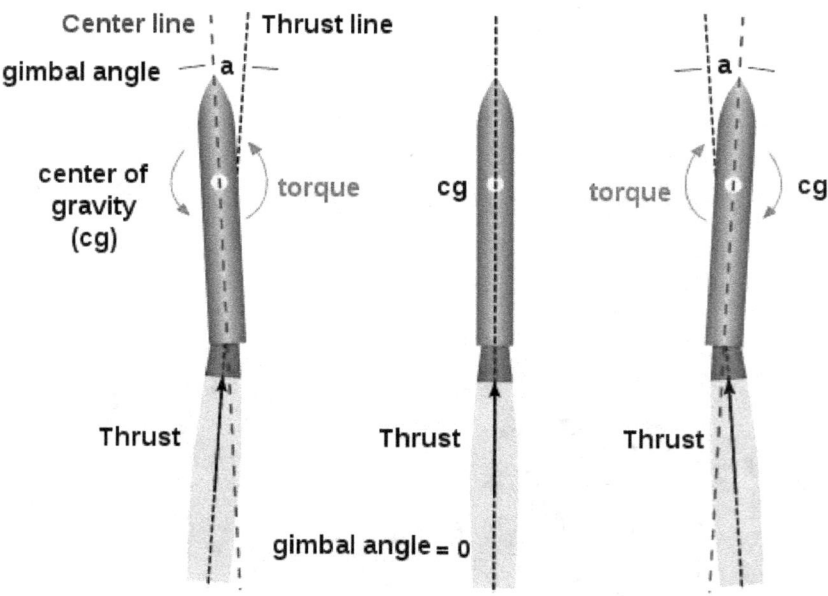

Stages and sometimes boosters are ejected once spent and reduces the weight of the remaining vehicle. The final stage that inserts a satellite into Earth orbit or beyond is often known as the upper stage. These are sometimes built into the satellite itself or more often designed as the final 'throw away' part of the rocket.

In my teenage years, a friend and I did produce many multi-stage model rockets. The best combination did seem to be three stages as with Apollo; two stages seemed to be more reliable for us. We did try a six-stage rocket once but never got off the ground.

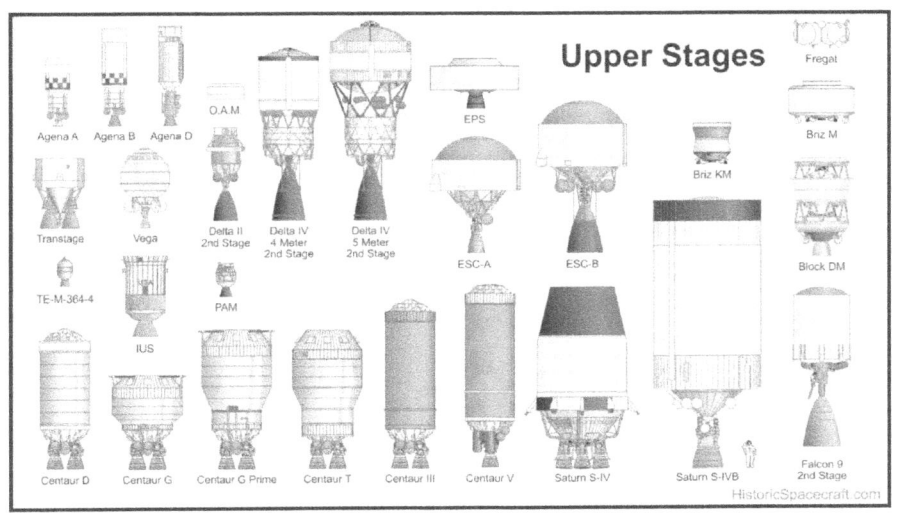

Above; examples of upper stages that put satellites into orbit or beyond.

Once in its temporary or final orbit, the solar panels (if included) must unfurl. These will generate power to charge the batteries before the pre-charged battery power runs out. Without this working, the satellite will become useless within hours.

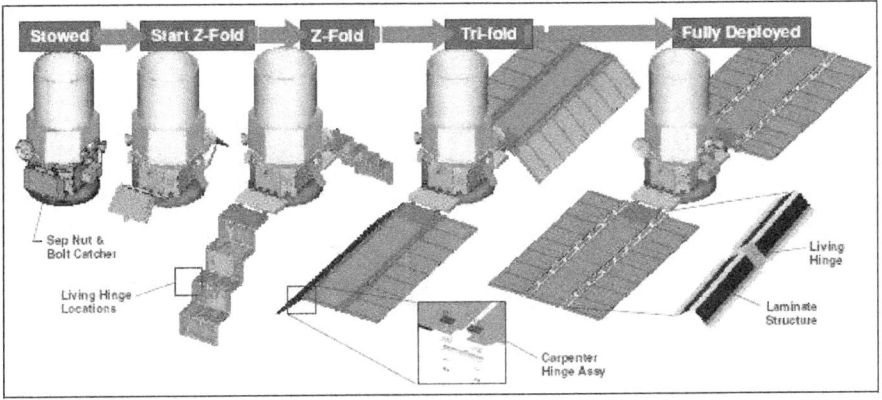

Some satellites will be parked in a temporary orbit to ensure all the systems are functioning correctly, recharge power supplies and link up with ground communications. The higher operational orbit may have some very precise positioning requirements and so the last burn is held back until all the conditions are met. The engine is ignited one last time. This is known as the Hohmann Transfer Orbit (marked thickly below).

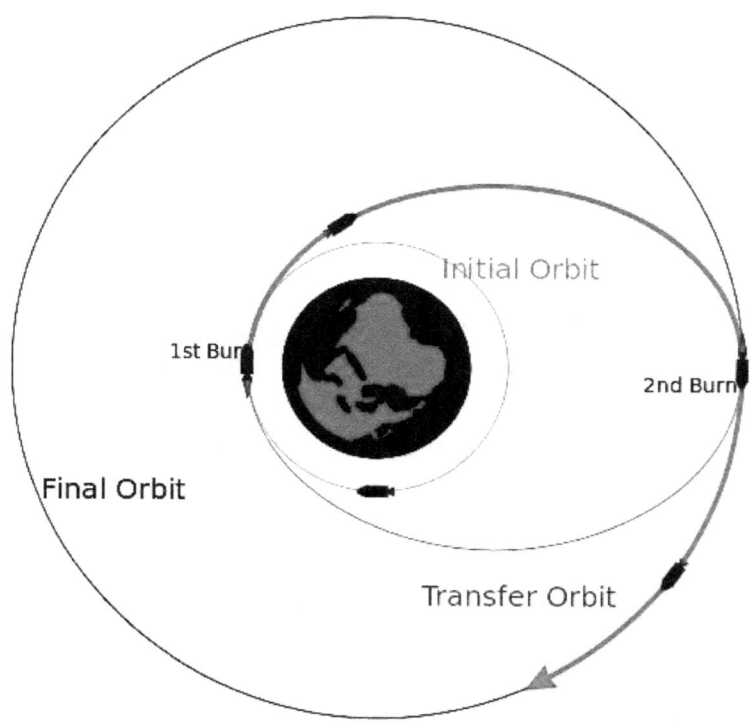

The same type of transfer orbit is used for destinations such as the Moon or mars etc.

Chapter 15   Ground Track

As the Earth rotates, the satellite is revolving around too. If it is not in an Equatorial Orbit, then gradually the satellite will pass over different parts of the globe each time it gets back to the starting point and begin another circuit.

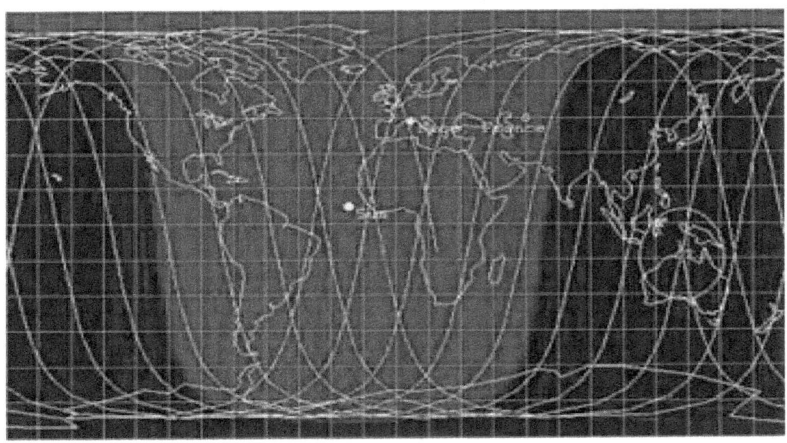

After many orbits, the ground track roughly repeats itself. Most satellites travel roughly West to East; with the earth's rotation. This example is of a Polar Orbiting Satellite.

Many satellites can be viewed twice in one evening from around April to August in the Northern Hemisphere and August to April in the Southern. The same satellite can pass over again after 100 minutes or so but as the Earth has rotated by 15° during that time, it will appear in a different part of the sky.

The ground track is defined by imagining a straight line from the satellite to the earth's centre; where it reaches the earth, this is the moving ground track.

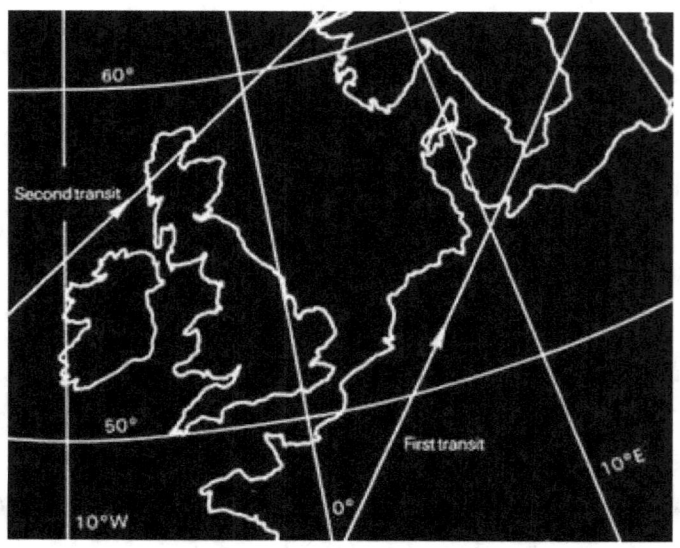

This example satellite can be seen throughout the UK twice. If the satellite was 500 km up and the observer was in London for example, the first pass would be around 60° up to the south and the second 35° up toward the north. Both will travel roughly west to east.

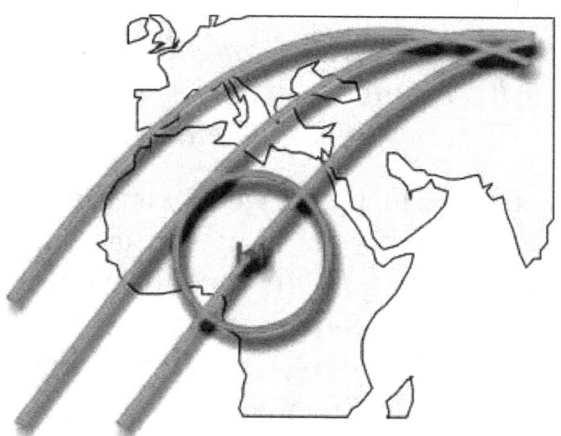

The previous diagram shows the ground track of the International Space Station on three typical orbits.

Every year the station is pushed into a higher orbit as it slowly falls due to thin atmospheric drag. The timing of each pass then changes. The altitude can drop by as much as 50 km a year; without such station keeping it will fall out of orbit as Skylab did in 1977.

Probably the most unusual ground track is on a highly elliptical orbit. As the satellite increases its altitude, it will physically slow down. If it is high enough, the earth's spin can overtake the speed of the satellite and make it appear to travel backwards in the sky for the Apogee (peak) part of its orbit.

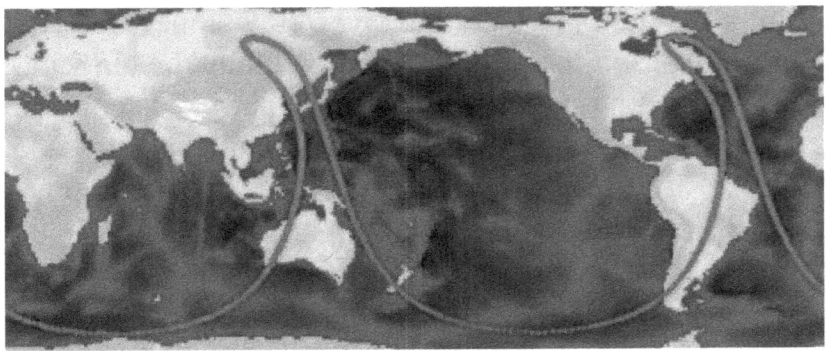

Chapter 16   Satellite Types

Research satellites; these measure properties of outer space such as magnetic fields, cosmic rays, micrometeorites and other objects that are difficult to observe from the earth. Examples include observatories designed to study radiation from the Sun, light and radio emissions from distant stars and the earth's atmosphere. The most well-known are the Hubble Space Telescope and the Compton Gamma-Ray Observatory.

Communication Satellites; these provide a worldwide linkup of radio, telephone and television. The first communications satellite was Echo 1; launched in 1960, which was a large metallic balloon that reflected radio signals. Relay 1 and Telstar 1, both launched in 1962, were the first true communications satellites; Telstar 1 relayed the first live television broadcast across the Atlantic Ocean.

A network of 29 Intelsat satellites in geo-synchronous orbit now provides instantaneous communications throughout the world.

Weather / Meteorological Satellites; provide continuous, up-to-date information about large-scale atmospheric conditions such as cloud cover and temperature profiles. Tiros 1, the first such satellite was launched in 1960, it transmitted infrared television pictures of cloud cover and was able to detect the development of hurricanes and to chart their paths.

The Tiros series was followed by the Nimbus series, they carried six cameras for more detailed scanning, and the ITOS series was able to transmit night photographs.

The Nimbus Series were named after the Latin word for Cloud. Seven were launched from 1964 to 1978 and includes one launch failure. These were originally designed as weather satellites but as other instruments were added, the role expanded and became weather + Earth observation satellites. Revolutionary instruments included Ozone gas imagers, Coastal Zone Colour Scanner, Microwave and Infra-Red Radiometers.

There is a clip that describes the history of the Nimbus series of weather / Earth observation satellites on the Satellites Classes page on www.outerspacebooks.com

Other weather satellites include the Geo-Stationary Operational Environmental Satellites (GOES), which send weather data and pictures that cover United States; China, Japan and India.

This view shows the potential coverage of a geo stationary satellite for weather forecasting or communications above the equator. This posture remains constant for years and would serve Canada, USA, Mexico and northern South America.

Navigation Satellites; developed primarily to satisfy the need for a system that nuclear submarines could use to update their own inertial navigation. This led the U.S. navy to establish the Transit program in 1958; the system became operational in 1962 after the launch of Transit 5A. This provided a constant signal by which aircraft and ships could determine their positions within a kilometre. However, the Transit system had a severe limit; non-global coverage. There were always large areas of the globe that was not covered.

Today, Navigation Satellite for Time and Ranging/Global Positioning System (Navstar/GPS) consists of 24 satellites approximately 11,200 miles above the earth. It provides greater accuracy in a shorter time 24 hours a day to a position accurate to within 2 metres. Because of other improvements such as memory cards, the GPS system now has equipment that is more compact. The American Navstar is the most used such system today.

The new European Galileo GPS system; 10 times more accurate than the original US NAVSTAR system. These are still being launched as of 2019 and not yet operational.

Beidou is China's satellite-based navigation and global positioning system. It began operations is 2011 with 10 satellites, succeeding an experimental system that became operational in 2001. They plan for a constellation of 35 satellites when completed in 2020.

How does GPS work?

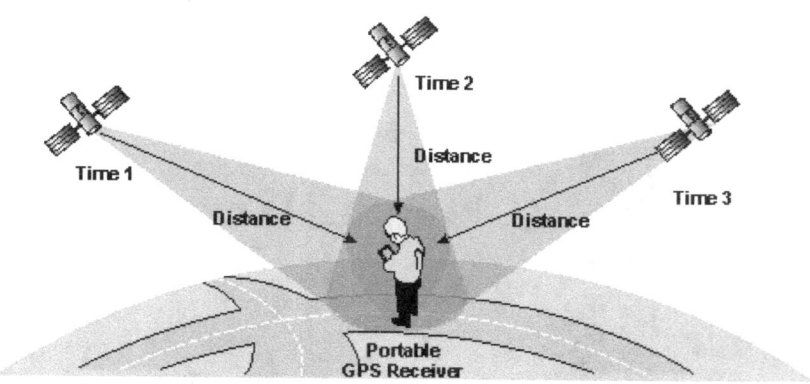

Applications satellites; are designed to test ways of improving satellite technology itself. Areas of concern include structure, instrumentation, controls, power supplies and telemetry for future communications, meteorological and navigation satellites.

Military Satellites; have been used for a number of purposes, including heat camera that track missile launches; sensors that listens on secret conversations; and optical cameras for photographing bases. The latest machine that catches the imagination is the X37B; a top-secret reusable unmanned shuttle launched from Vandenberg Air Force Base, California.

This incredible vehicle is a mini space shuttle without astronauts. It is reusable and launched inside an Atlas rocket. It can stay up for more than a year and land on a runway. It contains secret equipment. This is top secret... Well not quite...
www.outerspacebooks.com ; 'war in space' page.

The next image is of a back-up satellite called SOLRAD / GRAB that was to officially study the solar radiation in Earth orbit. Its true secret role was to detect Soviet air-defence radar positions.

The next image is of a capsule officially titled as a Transport Supply Spacecraft. It was launched and docked to the Salyut 7 space station in 1983 and returned five months later. It was a trial spy satellite.

Science Fiction stories have often touched on Killer Satellites & laser beam weapons. Even though they seem far-fetched, some of these have actually been tried in space with some success.

Watching satellites by amateurs and professionals is one of the very reasons why this is an important subject. It ensures global agreements of keeping weapons out of space are honoured; another example of how this harmless hobby can change the world.

Remote Sensing;
The USA has launched a series of Landsat remote-imaging satellites to survey the earth's resources by means of special television cameras. The data from remote-imaging satellites has also been used in archaeology. Russia and other nations have also launched such satellites; the French SPOT satellites also provide detailed photographs of the earth.

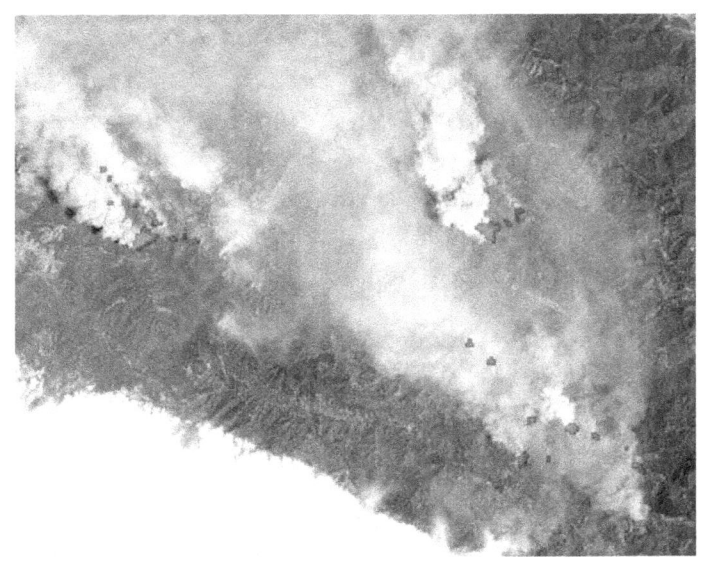

A Landsat 7 image of fires in California. The source of each fire is clearly marked to aid fire fighters to tackle the blaze itself rather than wasting time searching through the smoke.

Britain's 'Big Freeze' of January 2010 by Landsat 7. Taken on a rare clear day.

The latest data shows the northern jet stream has slowed due to the lessening temperature difference between the arctic and lower latitudes.

NOAA image

Ground Tracking Satellites

Tags are often used on road vehicles, ships, aircraft and even animals. A stolen vehicle can automatically trigger a transmitter to a satellite and back to a ground station. The police are informed and can easily locate the vehicle. Such devices are placed in a secret compartment onboard so the thieves cannot locate and remove it. Criminals are tagged with the existing GPS system. Aircraft have transponders and so do large ships for location and navigation assistance.

Creatures such as whales, bears, the big cats, sea turtles, sharks and eagles are often tagged to study migration patterns etc. Since 1978, several satellites include an on board system called ARGOS, this is devoted to animal such tracking.

Bio-Transmitters must be less than 5% of the weight by law on the creature it is attached to. This ensures that it does not spoil its natural lifecycle. Data by the device is gathered about the animal's position via GPS, but also the depth underwater through a water pressure sensor, or perhaps an altitude air pressure sensor in the case of birds.

Each Bio-Transmitter has an on-board battery and is designed to record and transmit information direct to the Argos related satellites. On land animals and birds, some sensors have a solar cell to recharge the battery and extent the life. This data is normally stored and transmitted once a month or so. The strap is often design to fall off the creature shortly after a while.

Gathering such information helps with predictions of the spread of diseases such as Avian Flu.

In 2017, a new system was installed on board the International Space Station. Throughout the year, ISS covers around 78% of the earth's surface, ideal for receiving tracking data for most Bio-Transmitters. A company in Wareham, Hampshire, England produces a wide variety of Bio-trackers.

Anybody interested in this research into animal lifestyle discoveries should explore www.biotrack.co.uk – it is free.

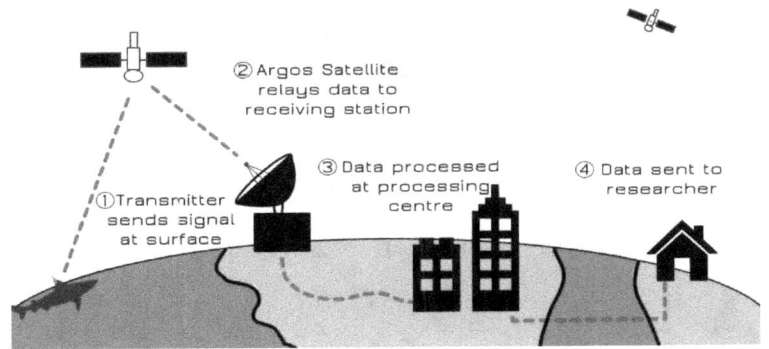

Chapter 17 Space Weather

Even though we generally class space as a vacuum; i.e. nothing in it, this is not strictly true. The Sun generates power in various forms that spreads out across the solar system via the Solar Wind. Even the probes Voyager 1 & 2 have detected some of these effects billions of miles beyond the orbit of Neptune.

These can alter the earth's upper atmosphere, magnetic field, electrical power grid, ground & space radio communications and satellite orbital drag. Some solar storms are so powerful they can burn out electronics in satellites and make them useless. The Sun goes through an 11-year long cycle of a minimum state of output to a maximum. 2014 was the last maximum.

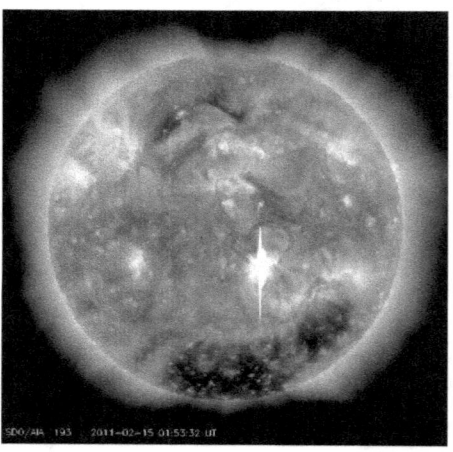

The bright spot near the centre of the Sun's disc in 2012 is a rare X-ray rich flare. It was aimed straight at the Earth and the energy from it was a serious threat to operational satellites.

Even minor disturbances in the upper atmosphere, the Ionosphere, can produce errors in the GPS signals. If a satellite navigational system in a car for instance is giving an incorrect location tens of metres out for a short period, then a minor magnetic storm from the Sun may be the cause. Charged plasma in the ionosphere will bend the satellite signal giving a false position reading. New GSP systems such as the European Galileo GPS uses two frequencies from separate time signals. This should eliminate such errors.

The Sun's magnetic activity pushes away and distorts the earth's magnetic field. Such an effect changes by the hour and requires 24hr monitoring. All satellites are within this region. Uplink / downlink communications can be affected if the distortion becomes extreme.

A series of satellites and probes blasted millions of miles into deep space monitor the Sun and all the effects it has on the earth's local environment. Some solar magnetic storms are powerful enough to destroy billions of dollars' worth of satellite hardware in orbit and may even endanger the lives of astronauts.

Electrons from solar storms can flow down toward the Earth and effect uplink and downlink satellite communications. These are mapped and operators on the ground using data can be warned of poor reception. Some disruptions may last a few minutes to a few hours.

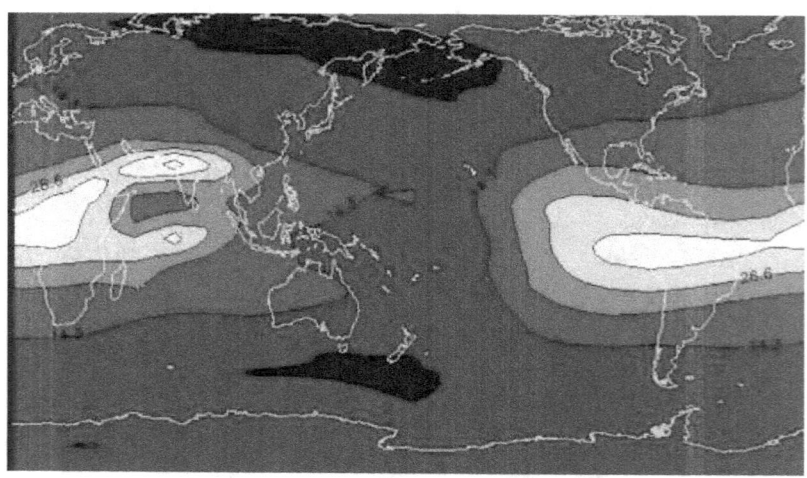

This is one such communication warning from NOAA in October 2014.

The Total Electron Content (TEC) is shown on the above map with intense areas marked in white and less troubled areas in dark blue. On this occasion, a journalist in northern India may not get a clear signal to a satellite for a BBC broadcast in England. Another journalist in New Zealand will get a prime spot on the news instead.

During very intense bursts, some satellites may be shut down completely and timed to wake up after the predicted event. Some have automatic systems detecting such dangers on a minute-by-minute basis.

The NOAA Space Weather Scales above were introduced as a way to communicate to the public the current and future space weather conditions and their possible effects on people and systems. This weather service by NOAA is a very young science and new prediction models are being tested constantly.

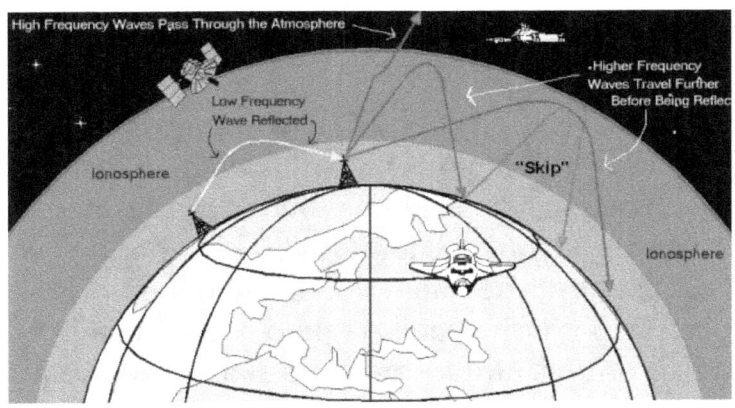

Solar activities change the Ionosphere and magnetic field of the earth. This alters the behaviour of long-distance ground and satellite links. Space weather due to the solar wind requires constant monitoring to keep our complex civilisation on track.

A solar eclipse reveals some of the Sun's activity – photo by the author Aug 2017, South Carolina.

Chapter 18 Space Junk

All kinds of junk exist in Earth orbit. Thousands of items including large rocket boosters, dead satellites, panels, tool kits, gloves and flakes of paint litter space. Satellites on rare occasions have collided and produced thousands of new pieces of rubbish.

The most worrying case of all was an anti-satellite weapon system tried out by China in January 2007. It destroyed an old satellite, called Fenyung-1C and created a massive expanding cloud of debris that is now a hazard to other satellites 800 km up. They were told off and promised not to do it again; but this problem will go for years. As time passes, this debris field will drop in altitude and eventually even cause a threat to the ISS; it will pass through the field twice per orbit.

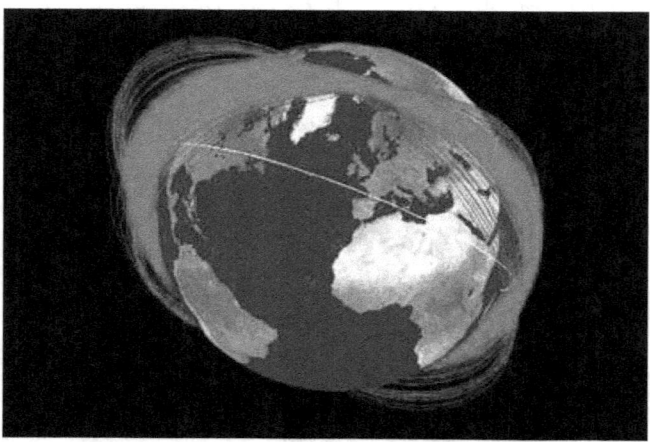

The debris cloud is tracked by radar and visually by North American Air Defence Command-(NORAD) to alert active satellites and sometimes have to change orbit to reduce the chance of a collision.

Kessler Syndrome

Donald J Kessler first wrote a depressing and dangerous idea in detail. He worked at the Johnson Space Center near Houston, Texas. He developed what is now known as the Kessler Syndrome; collisions become increasingly likely as the amount of space rubbish increases in orbit around the earth. Each collision creates more debris that can cause further collisions and so on. Kessler first published his ideas in 1978. This inspired the story behind the British hit movie 'Gravity.'

Early signs of this predicted that there would be an increase in unknown satellite failures. Most satellites that cease functioning do so for a known technical cause by studying the data on the satellite's condition. But unexpected failures are likely to be due to space junk hitting a vital part of a satellite. Europe's largest ever earth-study satellite, Envisat, ceased functioning for no known reason on 12 April 2012. It cost over £2 billion; it was built in Stevenage, England. Other similar failures have followed; all in low Earth orbit.

Another predicted sign will be collisions of complete satellites; the very first confirmed one was on 10 February 2009 between Iridium 33 and Cosmos 2251 over Siberia. Both were fully operational at the time. As they went off line at exactly the same time, it wasn't rocket science to imagine what may have happened. The two orbits were studied and realised the dreaded truth. Over 1,700 pieces are now tracked adding to the artificial cloud that surrounds our planet.

On February 15 2015, a satellite called DMSP-F13 experienced a sudden peak in temperature followed by a spin. Satellite spotters saw a large debris field near the satellite. This was almost certainly due to something hitting a fuel tank perhaps.

If the Kessler Syndrome takes place in its ultimate form, then this entire region of space will become unworkable for centuries until the upper earth's atmosphere naturally decays all this material on re-entry. The chain reaction will be unstoppable and alter our daily lives back to the 1960's. (Some may think of this as a good move).

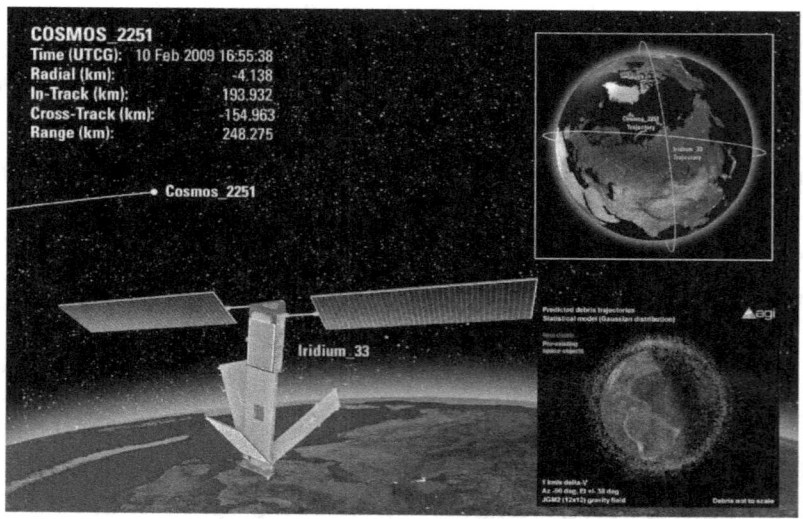

Possibly the largest risk of increasing the debris figures in orbit is through anti-satellite weapon tests (ASATS). As of 2019, four countries now have the capability of destroying a satellite up to at least 500 miles. The latest nation to join the ASAT club is India.

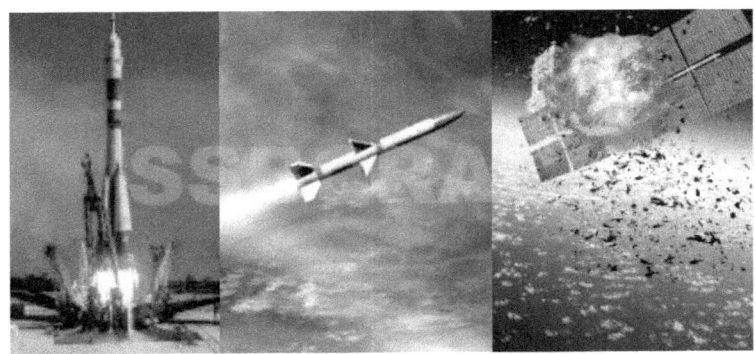

This debris cloud was predicted to have all entered the earth's atmosphere within weeks. After six months, much of it is still in orbit, spreading further out and posing an increasing risk of hitting an active satellite. None of this material can ever be gathered back or controlled.

The ISS is on constant alert from NORAD regarding a potential collision. Astronauts conduct several manoeuvres a year to reduce the chance of an impact. They are trained to respond to a call known as 'Conjunction Red.' This is where there is no time to calculate a burn to alter orbit as this can take a day or two of preparation. Instead, the crew evacuate to a Soyuz craft, power it up ready for an emergency undocking and then landing. If any object from the size of a cricket ball hit the Station, then it will almost certainly depressurise within seconds. The Space Station will 'unzip' at the weaker points and begin a chain reaction rather like a bursting balloon.

On a lighter note, the more interesting form of debris is spinning panels. These flat pieces that are tumbling end over end result in a 'flashing' satellite. This next example was the casing of an Indian satellite that passed over Arizona in August 2019. As it rotated out of control, the reflecting surface changed orientation and so produced a gentle fading in and out path across the sky. The star cluster on the left is called M45 in Taurus the Bull. *(Image by the author)*.

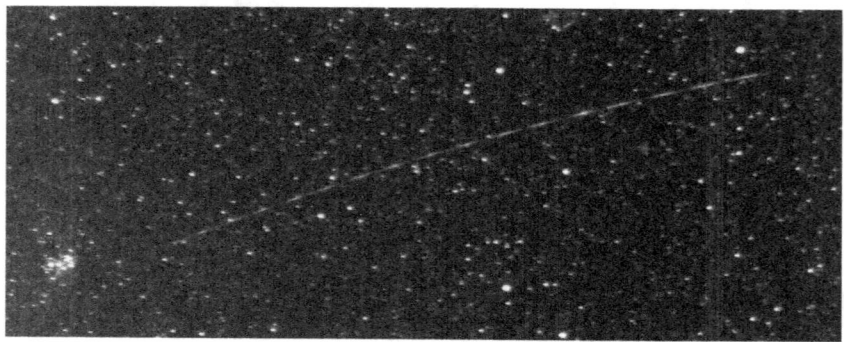

Some of the tougher form of debris will suffer from orbital decay, enter the atmosphere and land. Below is a US PAM module that landed in the Sahara and purchased by a space junk collector. It maybe about time we walked around with helmets on; although it would not have been a lot use with this one…

As of 2019, 28,000 pieces of debris are being tracked, most are larger than a cricket ball. It is also estimated that over 500,000 objects, too small to track, are also potential hazards that could easily destroy an operational satellite. They all orbit at speeds of around 22,000 km per hour. An estimated 100 million pieces of paint are also in orbit that could crack a window on a spacecraft.

A flake of paint in orbit hit the main antenna of the Hubble Space Telescope. At a collision speed of around 15,000 km per hour, much damage can occur and can make the craft useless. NASA image.

Huang Yijin and Zhang Hejin made a surprise discovery as they were walking through a forest near the village in China when they stumbled across an object that looked wildly out of place. They had no knowledge of satellite debris and concluded that it was a crashed UFO. They called the police who in turn called the China Space Agency.

The following 'UFO' turned out to be the nose cone of the Chang'e 4 rocket that launched a probe to the Moon just two days earlier on 20 May 2018. After a short stay in a local grain store, the nose cone was moved briefly to a museum before being reclaimed by workers from the launch centre.

By analysing the previous cases, it is clear the Kessler Syndrome has actually begun. If no further satellites are launched from this day on, collisions will still take place, and debris will increase from say 10 million destructive items, to a billion within a decade.

The British company, Surrey Satellite Technologies Ltd, is amongst the world leaders in this field. A system has been invented to approach a satellite, wrap itself around and this in turn increases drag

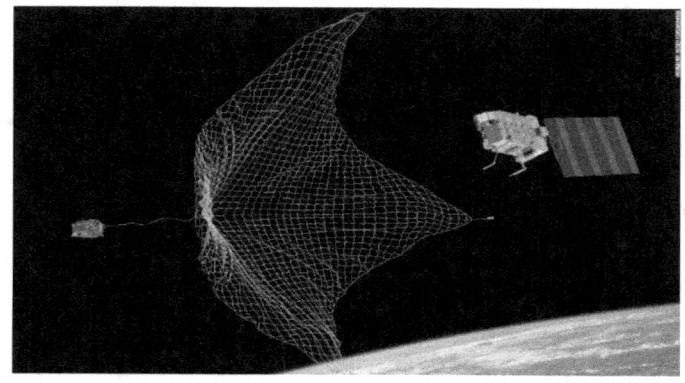

Artwork by David Duorosiesa

Chapter 19   How Bright? (Magnitudes)

How bright is a star, a planet or a satellite? Apparent Brightness is a measurement of how bright an object seems to be to a human eye regardless of distance. The lower the number, the brighter the object; it may sound back-to-front but that is the way it is. A very bright object such as the Moon can be minus something, a very faint object is always plus something.

All you need really to learn or at least appreciate are certain comparisons. Some of these following figures vary slightly over time as planets move around and some stars vary in brightness. Just keep referring to the next table to get the hang of it.

Object	Magnitude
The Sun	-26
The Moon	-12
Planet Venus	-4.5
Planet Jupiter	-2.5
Sirius – brightest star at night	-1.4
Rigel in Orion	+0.1
The star Vega	+0.03
The star Aldebaran	+0.8
North Star in the Little Bear (Polaris)	+1.4
The Dubhe in the Great Bear	+1.8
The star Kochab in the Little Bear	+2.2
The star Pherkad in the Little Bear	+3
Faintest naked eye stars	+6
Faintest stars seen through 10x50 binoculars	+9

This is a CubeSat; just 50cm x 60cm x 60 cm invisible to the human eye but just visible near the centre of this picture taken from Sittingbourne, Kent UK by the author.

Northern Hemisphere Observers

The next diagram is a simple map to show how to find the North Star (Polaris). Look for the 'Big Dipper.' In the UK, it is visible all year round. It is also known as the Plough, Saucepan or the Great Bear, Ursa Major. In the autumn and winter, it will be low down as in the picture below. During spring and summer, it will be virtually overhead. The shape will always remain the same, only its position will alter as the Earth spins and orbit the Sun. The stars Merak & Dubhe will always line up with the North Star or Polaris. True North is directly under it.

The four main guide stars for brightness comparisons can be;

Polaris Mag 1.4
Dubhe Mag 1.8
Kochab Mag 2.2
Pherkad at Mag 3

(Mag = Magnitude). These are ideal as the brightness are wide ranging, plus from a latitude above 35° or so, they are visible all year round. From lower down on the planet, the stars alter from one season to the next.

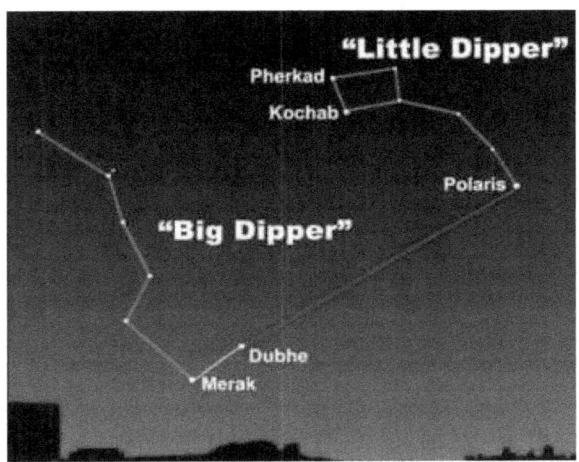

These stars can be used any night to compare the brightness of a passing satellite. If a predicted satellite is going to be Magnitude 4 (Sometimes written as Mag 4), you know that it is going to be a little fainter than Pherkad at Mag 3. If another is predicted to be Mag 1, then it will be a little brighter than Polaris. A guide to a satellite's brightness will enable you to make a decision on whether to observe it or not.

During the night, the Earth is still rotating; as the months pass we are also orbiting the Sun. This combined motion will change the position of the constellations but not the appearance.

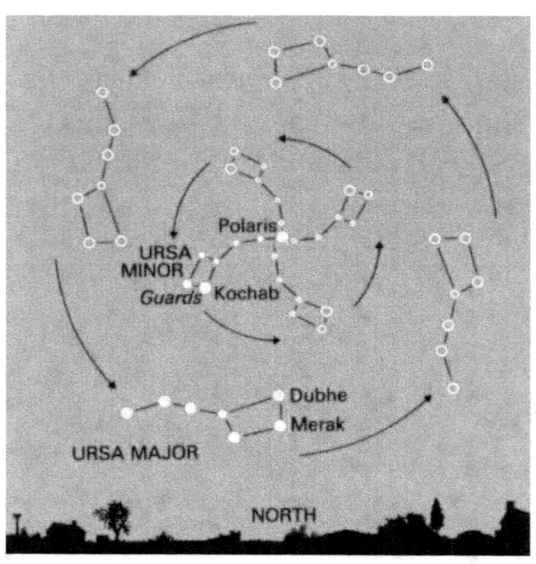

"The North Star (Polaris) is important but not very bright; like some people." President George Bush... I won't say a word.

Other conditions also change constantly. A large Moon may be in the sky and you may only just be able to make out Pherkad at Mag 3. If so and a satellite is going to be around Mag 4.5, then forget it. Go back inside and warm up. It will be too faint to see that night unless you use binoculars. The Moon orbits the Earth in a 29.5-day cycle of phases. Check a Moon calendar to see when it is visible in the morning instead of evening, and then the sky will be much darker to allow Mag 4.5 satellites to be witnessed.

Equatorial Observers
If you live less than 3000km north or south of the equator other guide stars will be required as a guide to brightness. Further South in New Zealand for example then there are stars that are visible all year as with the Great Bear in the UK.

Southern Hemisphere Observers

The next below is a good starting point for comparing the brightness of a satellite to a star. The Southern Cross is easy to detect, a small but bright constellation and visible all year from Australia, New Zealand, South America etc. The magnitudes are all marked against each star. Do not forget the higher the number, the fainter the object. Mag 6 is the maximum limit for the naked eye.

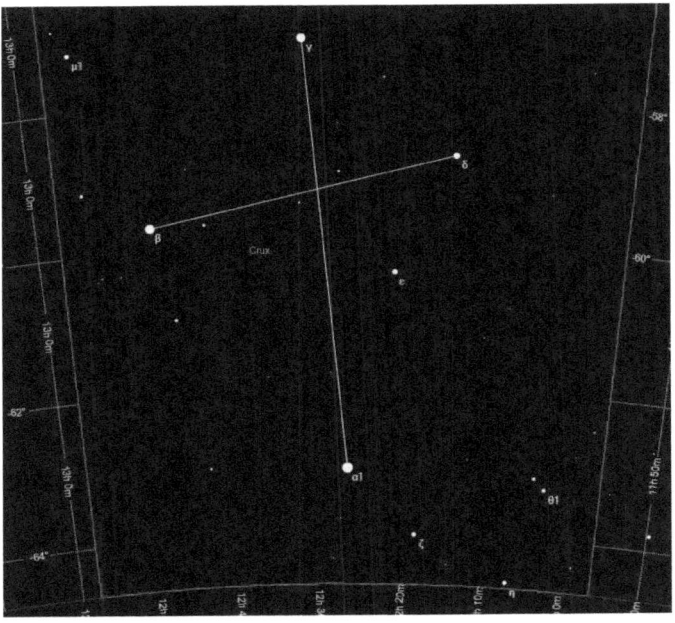

'Alpha Crux' is the brightest star shown at the bottom of the cross at Mag 0.8. Beta is to the left of the cross and is at Mag 1.3 and Delta to the right is Mag 3; a perfect group of stars for comparing satellite brightness.

Chapter 20   Lighting Problems

Several conditions limit how many satellites can be seen at any one time. These vary from night to night and place to place and are worth considering in advance to avoid being bored standing in the cold waiting for a moving dot.

Street Lighting
If you stood under a motion sensor security light that blinds your vision every time you make a move, then you are not going to be very successful in spotting any satellite or stars.

A general orange glow in the sky caused by town lights is called Light Pollution. A wet night will make it worse still by reflected light from the ground. Light Pollution will lower the faintest objects that can be seen.

Some parts of the UK are now having their local street lights turned off after midnight. This is to save resources, reduce pollution and money. The biggest reason in the UK is because of a lowering power supply due to coal station closures; to avoid a panic these other (good) reasons are publicly given. Once midnight arrives, the earth's shadow is directly overhead and no satellites will be seen. The only exceptional time of year will be six weeks either side of June 21st in the Northern hemisphere and six weeks either side of 21st December in the Southern. The shadow point will be away from overhead enough to allow a few low and all high satellites to be seen.

Moonlight

At least moonlight is very predictable. Do refer to lunar calendar tables (www.moonconnection.com) to show the phase for your chosen night of observing. It may be visible in the evening as a thin crescent. This would not cause any problem, but a full Moon or close to it will drown out all but the brightest satellites. After full moon, it will rise late, perhaps late enough not to cause a problem. From three days after full moon, it will not rise until after midnight. The Moon rises approximately 50 minutes later each day.

The following chart shows an example; the good dates for satellite observing are from the 8th to the 14th. Then again from the 27th till the end of the chart. All the other dates will be marred by too much moonlight.

Every month will be different as the cycle lasts 29 days 13 hrs – the average Lunar Month.

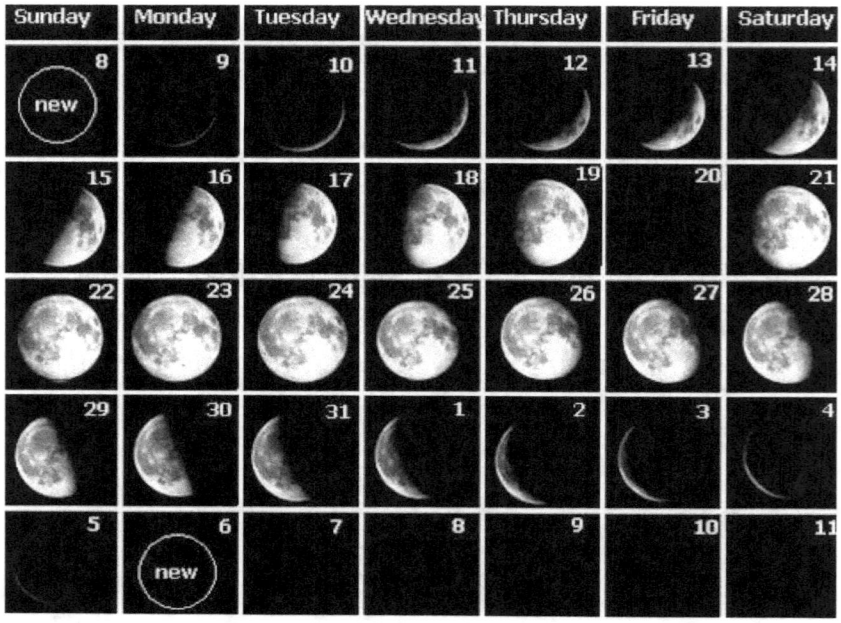

Look up the moon's phases for the month you wish to observe satellites. A little planning can save a lot of frustration. A large Moon in the sky will drown out all but the brightest of satellites. After full moon, it will rise later so as not to spoil the evening sky; approx 50 minutes later per night. (I do not know what happened on the 20$^{th;}$ perhaps it was cloudy)

<u>www.moonconnection.com</u>

Photography

A large Moon in the sky will also cause serious problems for photographing satellites. On an exposure of even just 5 seconds, the moonlight will completely ruin your picture. Save your time and effort for another night and just watch the brighter satellites pass instead. Darker skies are required for photography.

Clouds

To solve a cloudy night problem, just get the biggest fan you can find and... just kidding. There is not a lot that can be done about this, but one satellite in particular can still be seen through thin cloud – ISS. It is so bright that if it's due to pass high up, then at least part of the track can be observed. Look around, if a few bright stars that can be seen then you will not be wasting your time completely. Clouds do also reflect city lights and make the sky even less suitable. A few test photos beforehand will give an indication as to the limits of that night.

Even if there are some clouds around, they can make interesting shots and the brighter satellites can still be observed. On exposures of 30 seconds long for instance, moving clouds will be recorded as streaks; a pleasing effect.

Chapter 21   Visual Satellite Phenomena

They are many fascinating events to watch out for. Some of these occur every night, others are quite rare and very unpredictable; it is just a question of luck.

Entering and Leaving the Earth's Shadow
As a satellite travels into the earth's shadow, it fades for a few seconds as it starts to enter the shadow then vanishes completely. The time this takes depends upon the angle of approach. The satellite may head straight to the shadow at 90° and can fade in just 5 seconds. If it approaches from a shallow angle, it could take 30 seconds or more. On rare occasions, it may skim the edge of it and just make a fainter crossing of the whole sky without actually fading at all.

The angle of approach can be predicted by printing out the sky chart of a satellite crossing from www.heavens-above.com .

Very bright sightings such as ISS can appear pink. Imagine you are on the satellite; the sunset will be seen through the earth's atmosphere and red / pink light will get through but not the rest.

Iridium Flares
There is a constellation of satellites known as Iridium. It's a global communication system that includes satellite phones. They are at around 11,000 km up and normally can hardly be seen with the naked eye at Mag 4-7. But these have highly reflective gold coated antenna and reflect sunlight like a mirror. The following

picture is of a full size Iridium Satellite model at the Smithsonian Air & Space Museum, Washington DC.

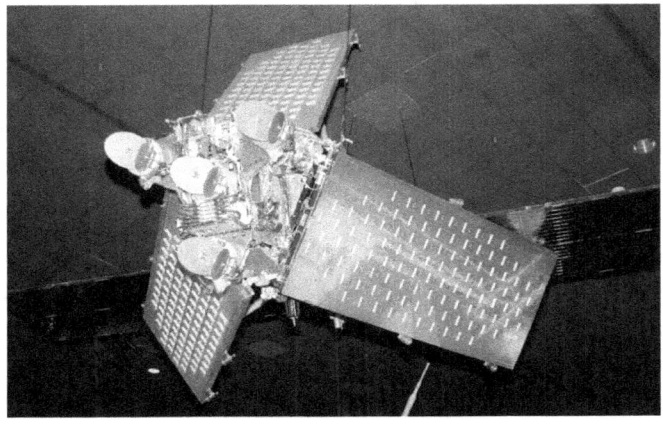

These can often be observed during the day as a flash of light lasting 3 seconds or more. At night, the flash can be seen for up to around 15 seconds. As they are all in polar orbit, this amazing phenomenon can be seen anywhere on Earth and have now become known as Iridium Flares.

I have recorded an Iridium Flare... Look up the page on 'Iridium Flares.'

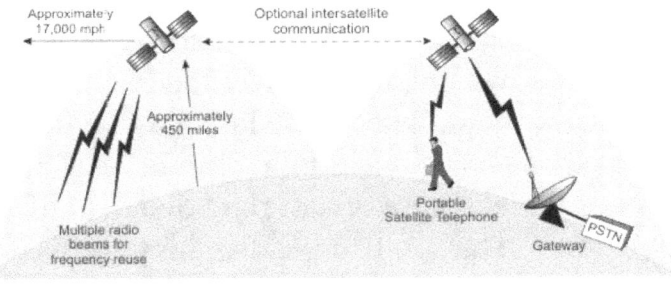

The Iridium constellation completes a modern global communications network.

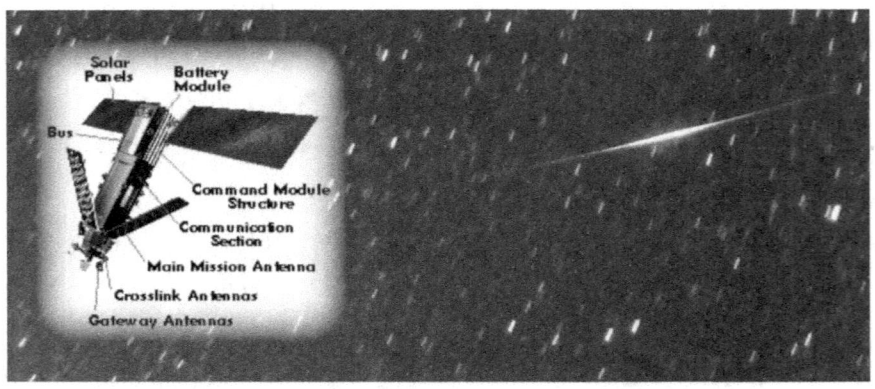

An Iridium Satellite can flare from being virtually invisible to become the brightest object in the sky within a few seconds. For details of how to view them, refer to the chapter called Heavens Above. These can even be witnessed in broad daylight.

If you find an Iridium flare occurring near you. Click on the 'ground track' map on the Heavens Above website to give an exact plot of the centre line. Travel to that location for the brightest view possible. Dozens of such events take place most months. Refer to Chapter on Heavens-Above for more details.

As from December 2018, these wonderful satellites are being brought down one by one over the Pacific Ocean, as they are simply outdated. The last may be down by the end of 2019. They are being replaced with the Iridium Next series; these do not flare due to a different design (below). They will be sadly missed, but other satellites do produce flares but not often as prominent as these do. *(Next – the replacement Iridium)*

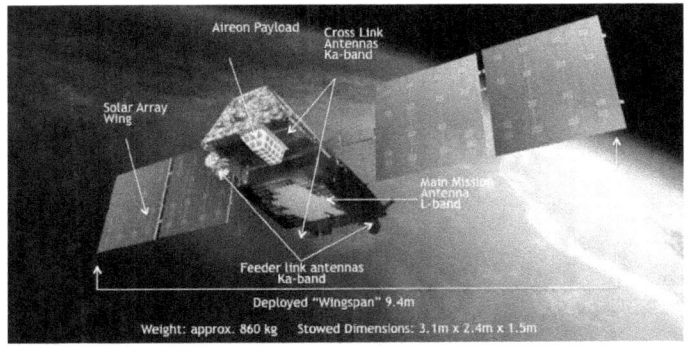

Flashing Satellites

There are spinning satellites that are uneven in surface area; elongated along one side. Most such examples are rocket casings that are cast adrift after launch. They are sometimes the final stage of a satellite launch that achieve orbit too. They can appear to flash bright to dim in a couple of seconds or very slowly taking a minute or more to rotate once. Some of these can be regular satellites where control has been lost, thrusters have fired to gain control but a failure of some kind has occurred. Nevertheless, most are simply discarded rocket boosters or satellite casings.

To record the length of a complete rotation accurately, chose a point in the flash such as the peak brightness, start a stopwatch with a finger, not by looking at it. Count five complete rotations as an example and stop the stopwatch at that peak in brightness. Simply divide the time by five. The longer the duration of timing will give you a more accurate reading.

The Synodic effect also includes short bright flares that are due to a flat surface on the satellite reflecting the Sun straight to you for a brief moment. It's also not connected to rotation but orientation of the Sun angle.

The picture above is of a slow rotating satellite that faded in and out every 6 seconds or so across the sky. It was around Magnitude 3. I took this in Devon August 2013; dark skies for faint objects.

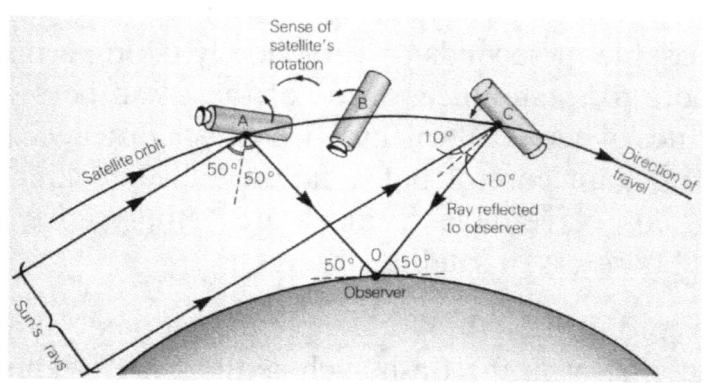

If an elongated satellite rotates, the amount of reflected light to an observer will change and alter the brightness. Some can become invisible to the naked eye then become visible again.

Dockings

You may be lucky enough to witness a docking or undocking event. A second fainter object will then be seen a few degrees away. The next image is such an example of a docking between the ISS and a Dragon supply vehicle.

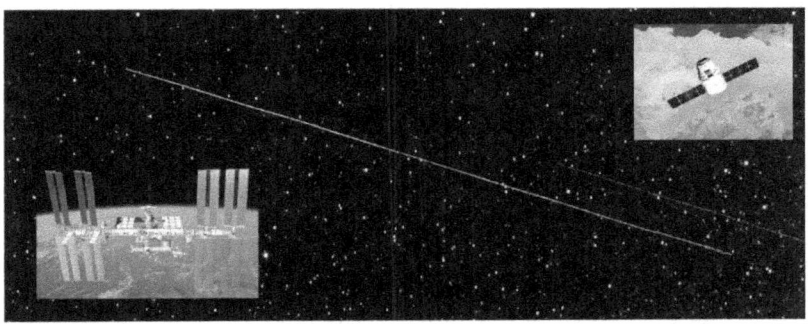

The very first docking I witnessed was in November 1973 when the last Apollo docked with the Skylab space station. As I saw it, both craft were just a degree or so apart. Skylab was about Mag 1 while the Apollo was behind it at about Mag 2.5. The associated Skylab / Apollo 4 capsule is now on display at the Smithsonian Air & Space Museum in Washington DC.

Fresh launch with trailing boosters
Whenever a new satellite is put into orbit, the outer casing or other rocket parts of the satellite often remains close by for a few days or weeks. Just look on the Heavens-Above website (Chapter 29) for pairs of objects that pass over at the same time and altitude in the sky. You should see a few every month.

The following picture was taken in November 2013 by the author. This is a Chinese military satellite called Yaogan Weixing that is using radar imaging in microwave. It was launched just four days before the picture was taken. Shhhh - it's all top secret.

The satellite is the top trail; the lower is the casing. (The camera did move slightly as the exposure began and produced a tiny tail on everything; ignore it please. I lost the infrared remote in the dark that night).

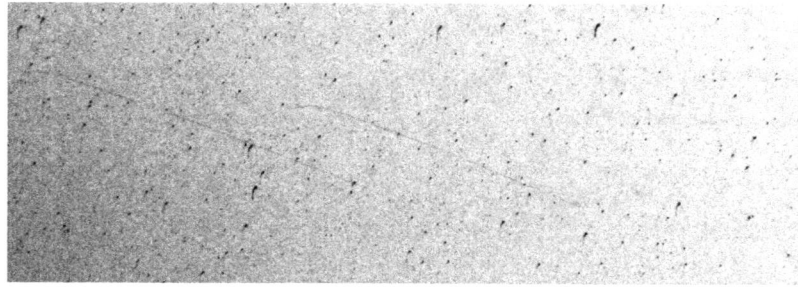

The top trail is the Yaogan Weixing satellite; it later tried to search for the missing plane Flight MH370 of April 2014. The Black & White negative version was to enhance the trails more.

Train of Satellites

A group of satellites in a similar orbit, all contributing data following each other is known as a Train of Satellites.

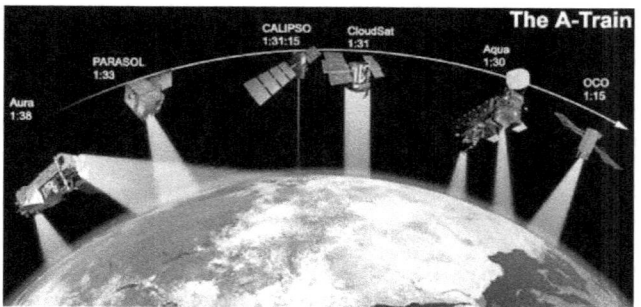

Formation Flying

Satellites can fly in formation of three in a triangular grouping. This is mainly for live 3D imaging. Landsat 7 flew as a pair with Earth Observing 1 and produced the first 3D hurricane data that led to new models of storm predictions.

The Swarm satellite formation are visible from anywhere on the globe. They are named Alpha, Bravo & Charlie, built in Stevenage UK and launched from Russia on the same rocket. This system adds to our knowledge of earthquake causes and predictions.

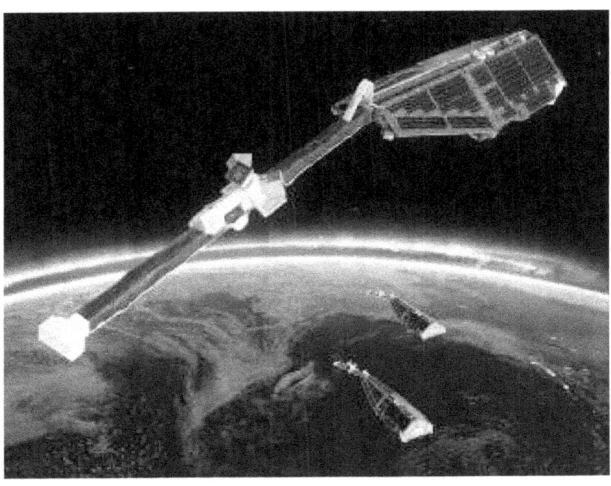

The satellite formation called Swarm. Designed to observe the ever-changing magnetic field of the earth.

Re-entry

All satellite orbits slowly drop in height due to atmospheric drag. We usually say that space is a vacuum; no air. The earth's atmosphere just gets thinner and thinner until there is no air at all. The vast majority is below 100 km. Even at 36,000 km up at geo-stationary orbit, a few molecules per cubic metre still exist. The weak impacting force will slow down the satellite and begin to drop. These machines will remain in orbit though for over 1 million years.

The Low Earth Orbit satellites are hitting millions of air molecules as it orbits at 7000 metres per second or so. For large objects such as ISS, it can drop from 400 km in altitude to 350 km in a year or so. An engine firing boosts it back up. Without such adjustments, it will simply fall out of the sky within ten years. This is precisely what happened to the Skylab space station and entered the atmosphere on its last orbit in 1979. Pieces were found across Australia.

An unknown satellite burning up in the atmosphere.

When a satellite is due to fall out of orbit, alerts appear on the Heavens-Above website – check out chapter *Which is which?* The last orbits are announced and you may be lucky enough to see it break up into many pieces and witness a spectacular event as it trails with smoke and debris across the sky.

The temperatures reached by re-entering debris follow a simple rule... 1m/sec per 1°C. Therefore, a satellite entering at 7000m/sec will rise to 7000°C. However, that only applies to the outside skin. Some parts can survive to the ground as demonstrated.

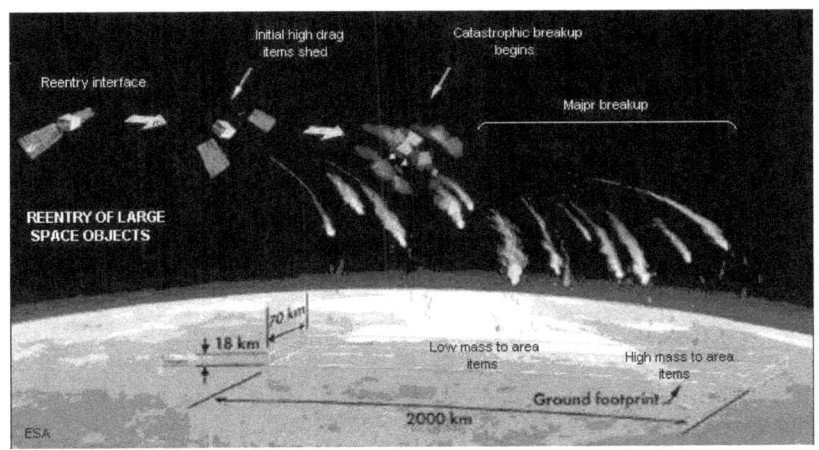

*Several clips are on the website;
'Satellite Re-entry' page.*

Dumping Fluids

When pee, fuels, or other liquids are dumped into space, they disperse into a massive cloud around the spacecraft. If the cloud is big and dense enough, it can be observed from earth.

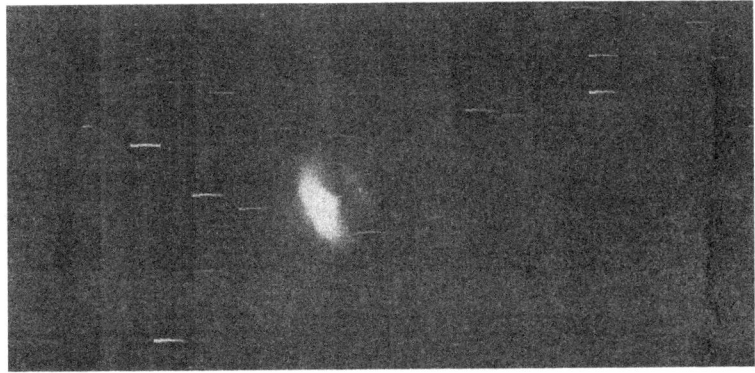

This fuel dump event was recorded from Spain during the Apollo 8 mission.

Chapter 22   Observing Techniques

Here are a few tips to enable a predicted satellite to be observed faster.

Dark Adaption

It does take a few minutes for anyone to get adapted to the dark. As soon as you step outside, your pupils will widen to allow extra light in. This takes around 5 minutes depending upon your age. For oldies like me, it can take 10 minutes to get fully used to the dark.

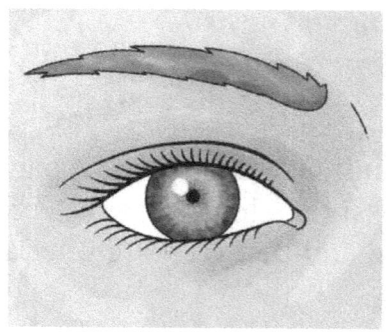 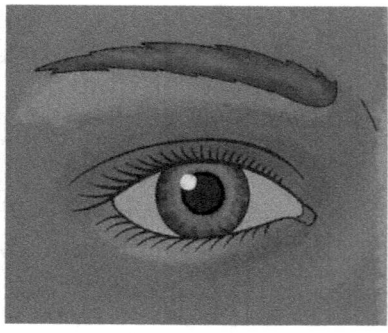

Naked Eye Scanning

When looking for a slow-moving object, keep your eyes fairly still in the general area where you expect to see a particular satellite; a moving object will stand out much faster rather than looking all over the sky randomly.

Binocular Scanning

For satellites below naked eye visibility, move the binoculars in a zigzag fashion along the predicted satellite track working backwards from where the satellite is predicted to be. Scanning at random means you will miss it; the satellite travelling at 7 km per second or so will not wait for you.

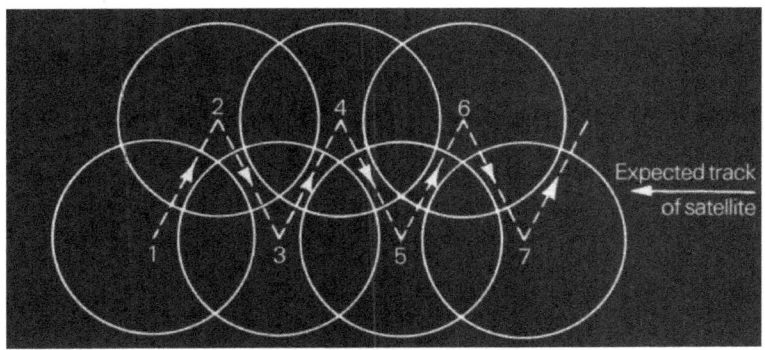

Geo-stationary Satellites

These simply do not move against the background sky. The Earth rotates minute by minute so the stars slowly change position. Use Heavens-Above (Chapter 29) to find out the compass bearing and altitude above the horizon for a group of these. Such objects can be viewed through large aperture binoculars and will remain there.

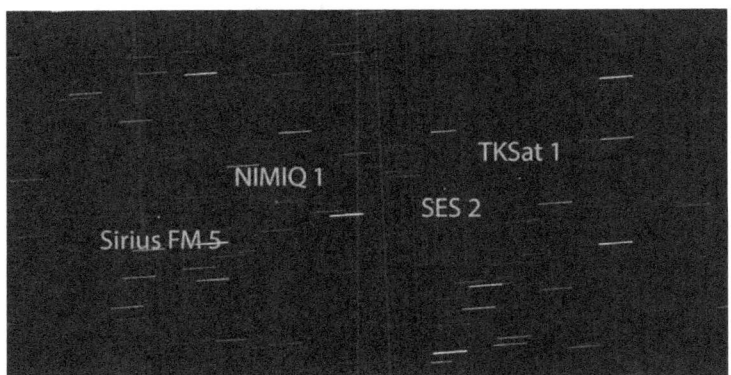

Use a five-minute exposure with a camera aimed at the 'Equator on the sky (Orion's Belt is a good guide as it's on it) and Geo-Stationary satellites easily reveal themselves virtually all night. The stars will streak as the Earth turns, but the satellites will not.

Chapter 23   Which Satellite is which?

The key to the entire hobby of Satellite Spotting is being able to tell one satellite from another. From the ground, we just witness a moving dot. To improve the interest in this object we have to know when they pass over, what it is, when it was launched, who built it and what it is doing. Other matters of interest include; will it go into shadow or is it tumbling to make it flash?

As such information changes by the hour, the only way to obtain this for your hometown is to use the Internet. There are several websites around including NASA for the bright satellites. I have listed a few below but have concentrated on one in particular. Heavens-Above seems to be the best site in the world for satellite spotting. It is completely free; you don't have to register in any way and does include thousands of objects bright, faint, live, dead, junk, low and high orbiting satellites of all countries.

Some of these websites are as follows…

www.n2yo.com A good site to see the live position of various satellites but full of adverts that cut into your experience.

http://spaceweather.com/flybys/country.php This one is good for starters. This is not very detailed though, no sky charts, but does include compass bearings and an indication of brightness. No information on Iridium Flares or other unusual phenomena. However, yes good if you do not wish to go into too much detail.

http://spotthestation.nasa.gov/sightings/ this one if for the Space Station only.

www.heavens-above.com This one is the business. The No.1 satellite spotting website for the beginner and experienced observer with full sky charts, compass bearings, brightness information, enter or exit shadow positions. Predictions can be sought for several days ahead or even in the past. You may have seen one in particular the previous night and you are curious to what it was.

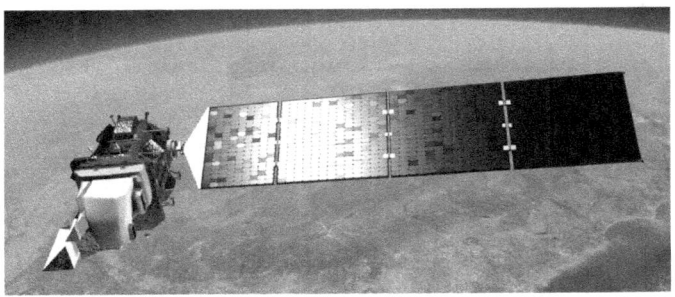

Landsat 8; knowing which satellite is passing over together with its function bring home the reality of how our modern civilisation operates.

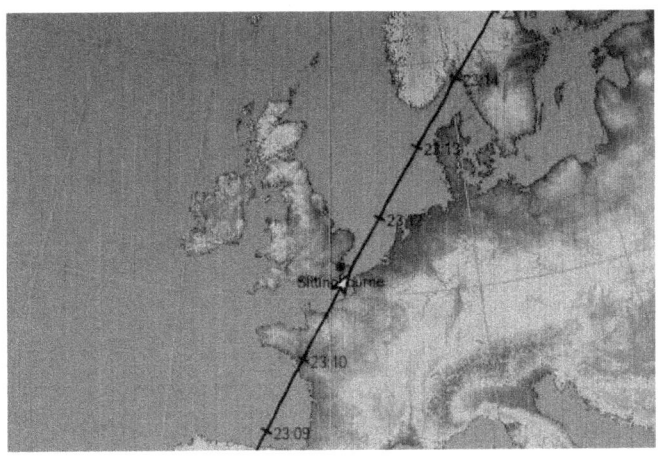

Heavens Above

The Heavens Above website really is the key to the hobby of Satellite Spotting; the World's most complete site available to the public for free. You do not need to register and is easy to use following the systematic guide in this chapter. Use the link below after reading this chapter and familiarise yourself with the site. www.heavens-above.com

All the satellite predictions are up-to-date. The further ahead you look in time though, the less accurate the forecast. If you intend to observe tonight or tomorrow; then there is no problem. If you want predictions four weeks ahead, then print it out but reprint the same night's forecast on the day if possible. Satellites can change orbit by firing a thruster, or through atmospheric drag if in a low orbit. One of them may have even been deliberately de-orbited and will never see it again.

If you saw a satellite the previous night, then the same information is available. Change the date on the predictions list and look up the time, direction and rough altitude in degrees above the horizon when it reached maximum height. The closest match should be the one you saw.

Register

Register a few details with Heavens Above, then you do not have to reset your location every time; just log in with a user name and password of your choice. You would also have the chance to post your own observations if you wish on unusual events or ask general questions to other users. You will not be bombarded with emails and adverts at any time.

Setting Location

After finding www.heavens-above.com the site starts assuming you are at 0° longitude and latitude. Click on Change your observing location…

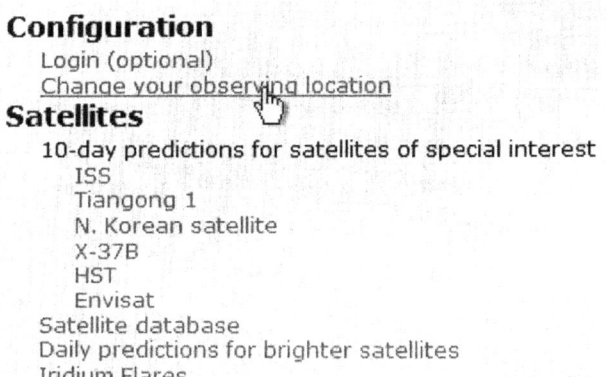

… Then type in your hometown in the top line and click search. Once you see your hometown accepted with a map, go to the bottom of the page.

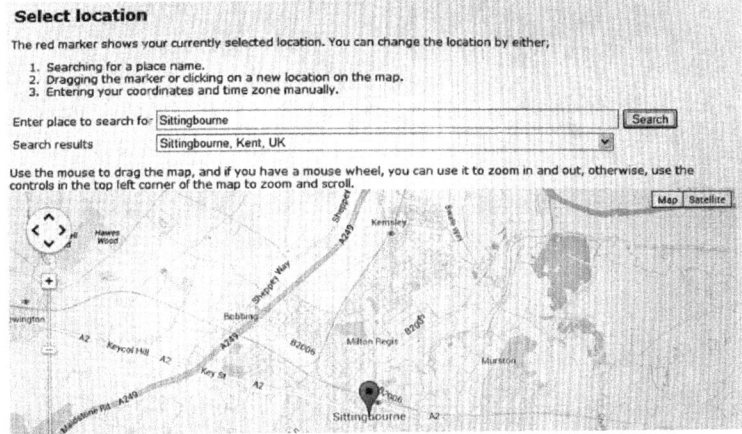

At the bottom of the page, you will find the 'Update' button. Click on it and the site returns to the home page, but from now on all the predictions are from your location. Regarding time zones and Summer Time / Daylight Savings, this is already taken into account. All the times for predictions are your own local time to avoid confusion.

Choose an option

From this point, you have the option of looking at a simple list of satellites visible that night. Click on the button as shown and a list of 50 – 150 satellites will appear as potential targets for you. The highest numbers will occur around March and September due to the Earth shadow angle and length of the night.

Satellites
 10-day predictions for satellites of special interest
  ISS
  Tiangong 1
  N. Korean satellite
  X-37B
  HST
  Envisat
 Satellite database
 Daily predictions for brighter satellites
 Iridium Flares
 Spacecraft escaping the Solar System
 Amateur Radio Satellites - All Passes
 Height of the ISS

This list could display over 200 satellites, so to reduce it, click on 'Minimum Brightness 3.5 and click on 'Update.'

Daily predictions for brighter satellites

Month May 2014 Day 17 ○ Morning ⦿ Evening [Update] [Reset to now]

Minimum brightness: ○ 3.0 ⦿ 3.5 ○ 4.0 ○ 4.5 ○ 5.0

Satellite	Brightness (mag)	Start Time	Start Altitude	Start Azimuth	Highest point Time	Highest point Altitude	Azi
Meteor 1-15 Rocket	4.4	21:18:04	10°	N	21:23:52	78°	E
H-2A R/B	2.6	21:19:37	10°	NNE	21:23:47	58°	E
H-2A R/B	3.0	21:22:17	10°	SSE	21:26:19	81°	W
SPOT 1/Viking Rocket	4.1	21:22:51	10°	S	21:27:58	61°	W
Cosmos 1939 Rocket	3.0	21:23:05	10°	SSE	21:27:16	88°	E
Gbstr 26 Del rocket	4.2	21:23:08	10°	W	21:29:13	84°	S
Cosmos 1943 Rocket	3.7	21:24:54	10°	SE	21:28:14	16°	

If you wish to change the date, the option is above; but not to far into the future as the predictions become less reliable.

The easier satellites to observe are those that fly almost overhead. Study the list of 'Highest Point' figures; the closer to 90° the better as this is overhead (the Zenith). The lowest I normally try to observe around 35° high. I chose one example next – **Cosmos 1943**, launched in 1988. It was predicted to be Mag 2.1 and pass at 87° above the horizon at 23:11pm.

CARTOSAT-1	4.0	23:05:15	21°	S	23:07:56	55°	W	23:12:12	10°	NNW
Cosmos 1656 Rocket	3.8	23:05:36	20°	S	23:09:15	54°	ESE	23:14:37	10°	NE
Cosmos 1943 Rocket	2.1	23:05:37	10°	SSW	23:11:11	87°	ESE	23:16:45	10°	NNE
ASTRO F (AKARI)	4.4	23:08:41	35°	SSE	23:10:02	81°	ENE	23:13:22	10°	NNW
CZ-4C R/B	3.9	23:09:57	23°	SSW	23:12:03	47°	W	23:15:38	10°	NNW

After clicking on the time, a full sky chart will appear with the satellite track marked on it with direction of travel and times.

115

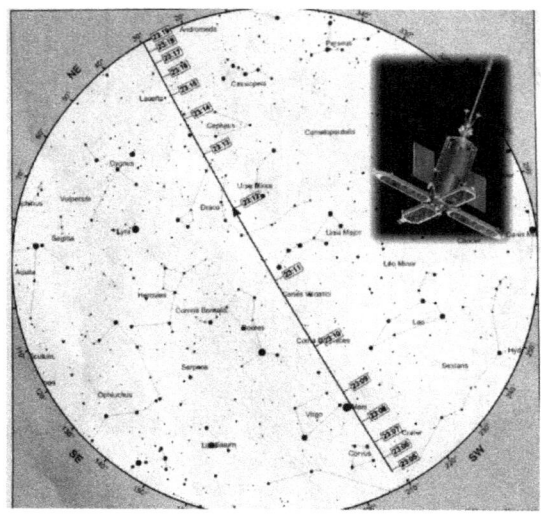

Go outside 10 minutes before the predicted maximum altitude of **23:11** in this case and get dark-adapted. Look for the overhead stars and see if you recognise them. It does not matter too much if you do not, the moving satellite should be seen anyway. In this particular case, the satellite was a rotating cylinder. This means that the observed brightness may be completely different to prediction. Brightness of an operational satellite is accurate.

To increase level of interest and home to you what you are actually seeing, click on the satellite name itself and details of the launch etc. will be shown.

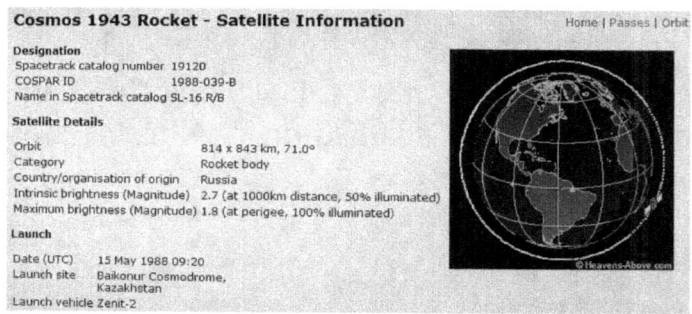

116

If you wish to see the satellite again, click on 'Passes' or more detail of the orbit use 'Orbit', it's a very well laid out website as you can see already. After a little practice, you won't need this guide any further.

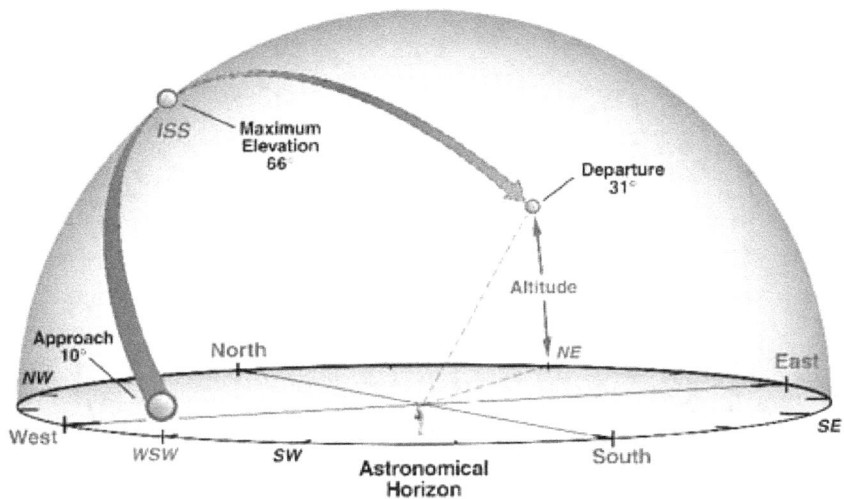

Above; if you do not wish to learn the constellations, then all you really need are compass bearings and an understanding of elevation in degrees. 0° is the horizon, 45° is half way up and 90° is overhead. All these details are given on Heavens Above.

Next is a close up of the actual event I observed. As the object was spinning, the amount of surface reflecting sunlight was changing second by second. It was invisible to my eye to start with then brightened, then faded again. It repeated itself every 20 seconds or so. This object can now be classed as a 'flashing satellite.'

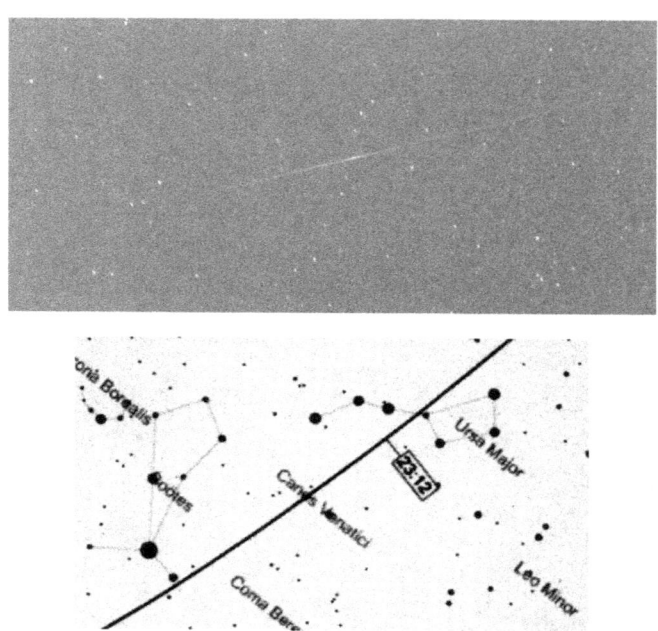

The predicted path of H-2A; US rocket booster 19th May 2014.

The next photo is of the actual satellite H-2A. The track cut short due to a 15-second time exposure. A longer exposure under the conditions that night would have fogged out the picture as there was a little heat mist reflecting local streetlights. Compare this against the predicted track.

Chapter 24 Satellite Track Photography

The simplest form of recording satellites is to photograph them crossing the night sky. With exposures shorter that 50 seconds, the stars remain virtually as dots while the rapid moving satellite will trail within the field of view.

My very first picture was taken in 1977 of the then abandoned US Skylab space station. The Daily Mail newspaper in the 1970 used to show the moon's phase, sunrise & sunset time. Whenever a bright satellite was visible, the time of the crossing would be included too. This exposure was around five minutes long; the stars have trailed due to the earth's rotation.

This was taken on an old Russian Zenith E camera with 400 ISO B&W HP5 film. I did not have a tripod at the time, so I laid it back on Mum's trelliswork for her roses and pointed the camera straight up. It was just steady enough without wind to buffet it during the exposure.

To focus such a camera on infinity was very easy; just rotate the focus ring as far as it will go anti-clockwise. Digital is not always so easy, but other advantages far outweigh the disadvantages. (I did knock off a rose; do not tell Mum).

Settings Required

It is important to be very familiar with a camera for any night sky photography. Do use the instruction book rather than guessing as I have done many times.

Regarding memory cards, the 45MB/s should be the minimum. This reduces battery consumption and wasted time between images.

With any Digital SLR camera, the following settings must be understood.

Field of View

Most cameras come with a Standard Lens; or at least a lens that can vary from wide angle, through to a slight zoom. A wide-angle setting will obviously allow a larger part of the satellite trail to be recorded. As the Earth spins, the stars will trail on the picture and become short streaks, but on a wide-angle setting, this effect is minimised. On exposures of less than 40 seconds, the stars should remain as points of light rather than streaks. Only the moving satellite will produce the trail you want.

ISO (International Standards Organisation)

This is a measure of how sensitive the camera is to light. The higher the number, the more sensitive it is, the fainter objects will be recorded. When film was used, the most common ISO available was 400. A reasonable mid-range priced digital camera often goes up to 12,000 ISO; perfect for night sky photography. The top cameras can now go up to around 400,000.

Refer to the Manual to set ISO. If its set too sensitive and you have some light pollution, then your image may fog over with bright orange glow and ruin the picture. Experiments by photographing the stars before the satellite crossing would be very helpful. ISO 1600 is a safe setting for an average sky and will record a moving satellite down to Mag 4 or so. ISO 3200 will record fainter satellites down to around Mag 5. Ideal if you have a dark sky and little or no moonlight.

You can try higher settings, but these only work generally with very dark skies where the Milky Way can be seen. A series of experiments on the night will help you decide. Try each ISO setting and expose for say 30 seconds each time. Look at the results on the LCD screen. After a little experience, you should be able to judge yourself what ISO setting you can get away with under certain conditions. My camera, Canon 600D & the 1300D can be set as high as ISO 12,800. The image does get 'noisy' though with lots of background rubbish appearing. Such images can be cleaned up with a little processing.

The next table is be used as a guide for minimum required setting can be used for moving satellites. As the sensitivity increases (ISO) so does picking up stray unwanted light from light pollution, mist, moonlight etc. The higher ISO settings require darker skies. This table is a guide only. The slower moving satellites (at higher altitudes) may be imaged more easily as the light will build up quicker on the chip.

ISO Setting (used to be known as ASA)	Faintest Satellite that can be recorded in Magnitude
200	+1
400	+2
800	+3
1600	+4
3200	+5

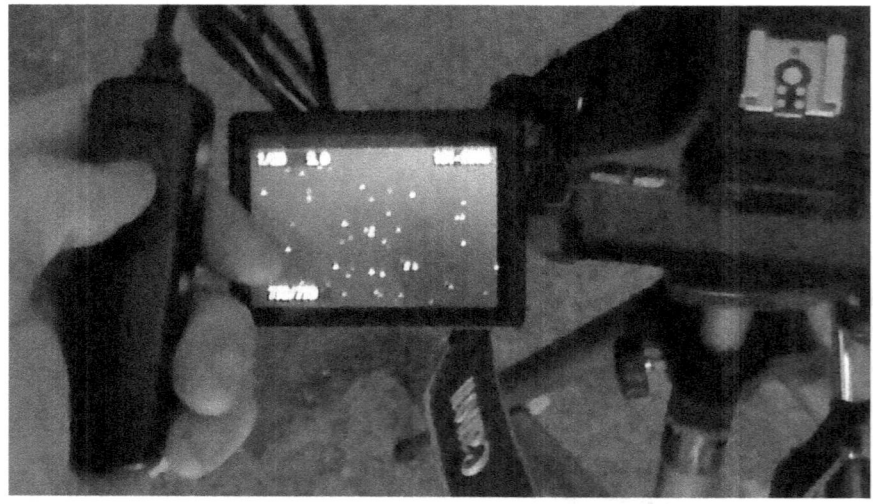

*If once displayed on a computer screen, images are still slightly out of focus (as shown here) it is possible to re-sharpen it by using software such as **Avanquest – InPixio Photo Focus**. It is not expensive and can save many photos; not just astronomical shots.*

Shot Alignment
The stars do not show very well in the LCD screen when night photography is required. A green laser mounted

on the camera flash unit will automatically be roughly aligned to the centre of view. Use gaffer tape to hold it in position. This can now be fired into the night sky and used as a guide to where the camera is pointing. Once in the correct position for your approaching satellite, turn off the laser. I have left in on in the next picture as a demonstration.

Exposure time

A typical satellite will cross the sky in around 5 minutes. The standard lens itself does not cover the whole sky so a satellite could travel from one side of the image to the other in around 1min 30 seconds depending upon the altitude above the horizon and the lens used. Therefore, a 1min 50 second exposure should capture a full trail right across the entire picture if you begin the exposure just before the satellite enters the cameras' view

As the Earth spins, the stars will rotate around the sky. They can create a pleasing effect if the exposure is long enough. Only the very darkest skies will allow this to work. The next image is a three-minute exposure and the stars have trailed a little. A meteor crossed the field of view during this time by coincidence. Light pollution from a nearby town is also building up; a little will not spoil the shot too much.

If the sky is not that dark, then play it safe and keep the exposure to no more than 30 seconds and the ISO setting no higher than 1600. You will capture at least part of the crossing and numerous stars but not too much of the light pollution.

Remote Triggers
To avoid a jerky movement on the camera as an exposure commences, it is best to use a remote trigger device, it should only cost around £2-5. If you don't have one, just put a black cloth gently over the lens; begin the exposure and pull away the cloth (often known as 'the black hat trick'). A two second delay in exposing the sensor to light will allow the vibration to die down before full exposure to light begins.

Remote triggers come in two forms; infra-red remote and a hard wired cable with hand button; the latter tends to be more reliable for astrophotography as they do not require batteries that are often affected by the cold or damp conditions. The camera can be set on say 30 seconds; by pressing the button on the remote, you commence the exposure without jerking the camera. Gently let go the remote, put your hands in your pockets, and wait. If you set the camera to 'Bulb' exposure time, this is a full manual timer. When ready for the picture, just press the button on the remote; as long as you have it pressed, the camera will continue to expose. Wear gloves if it is cold.

A cable trigger is recommended over an infrared trigger as you will never lose this in the dark and does not require an extra battery that can fail without notice. Many batteries hate cold temperatures. Tie up the excess cable to avoid it trailing on the ground or becoming entangled with the tripod legs. Satellites will not wait for you to sort yourself out before a scheduled appearance. £8 ($10) should do it.

If you want to be flash, you may wish to opt for a Digital Remote These are only around £5 extra than full manual. This will enable you to time each exposure as long as you wish and keep your hands in your pockets rather than keeping a finger on the button. I still prefer to full manual remote.

Tripod

A quality tripod is essential. A used tripod should not be any more than £15 ($22) as a rough guide. Do not use tiny floppy weak tripods; use something that is designed for a heavy camera. The weak versions will slowly 'droop' during the exposure and make your starry pictures useless; all that time planning, experiencing perhaps freezing cold conditions... wasted.

Batteries

Ensure the battery in the camera is fully charged before the evening (or morning). Long exposures use up power very rapidly. The cold also has an effect and weakens the battery power. It is strongly advisable to purchase extra batteries as spares; most are no more than around £5 ($7) on ebay. I usually have four batteries on standby in my pocket. Once they are flat, keep them separate from the charged ones in a different pocket... not rocket science I know. Working in the dark, trying to capture satellites at a set time and position, require practice. They will not wait for you while you change a battery after accidentally putting a flat one back in.

Previous; Essential equipment for satellite trail photography includes a Digital SLR with standard or wide-angle lens, tripod, spare batteries and a remote or black cloth to place over the lens until ready for picture taking. If you have not a remote trigger, place the cloth over the lens, start the 30-second exposure or so then gently whip off the cover.

First Trial Run
If this is your very first go at astrophotography, try photographing a plane passing overhead at night. It will trail just like a satellite. When you see the picture on your computer screen, the stars should be pinpoints and

the plane should show a nice trail with all its flashing lights spaced out.

After an initial satisfying 'hooray' zoom in and see if it is still in focus.

The next image is an example of a tumbling satellite seen from Arizona, Aug 2014; a piece of debris rotating every 30 seconds or so. This picture recorded one complete rotation. As the object became sideways on, it reflected very little light, then as it rotated full on, the light increased. The slight 'jaggy' line is due to the point source of light moving from one row of pixels to the next on the chip. It is not the satellite waving around.

Kit Bag
Once you have produced a standard outfit for satellite track photography, it is good practice to find a bag for

the kit. This will save time, improve motivation to go out at a moment's notice and photograph perhaps an unusual event. Within a few minutes of deciding to produce an image, the data is there in the camera and you can upload your picture to a website such as Heavens-Above or even Facebook. The kit should include;

Camera / Lenses
Fast memory card (Video compatible / 45MB/s second minimum)
Remote Control / or black cloth to cover the lens
Tripod
Red & White LED Torches
Green or Blue Laser
Accurate Clock or watch
Spare Batteries for Camera and Laser + charger
Small Carpet or Rug to kneel on and reduce lost parts
Compass
Notebook and pen

Tip for pleasing result; use the white light torch on a tree or wall etc. within the field of view for a few seconds during the exposure. The object will be lit and the stars and satellite target will not be affected.

Chapter 25   Moving Image Recording

The secret to recording satellites in real time is obtaining equipment sensitive enough to detect and record them, plus accurate focussing. Many Auto focus systems have a hard time focussing on faint stars. A manual override is essential in every case.

Tips

When trying out a new instrument for the first time, practice tracking aircraft during the day. Align the camera so the aircraft is horizontal in the field of view and keep the same orientation as it passes over. Recording a satellite is similar. If the camera is kept horizontal, then its path seems to change as it orbits the earth. Then record bright stars or planets in the sky at night. Play them back on a TV to see how sharp the image is and how shaky. Build up experience on obtaining useful footage and then you will be ready to capture satellite passes.

Do not be tempted to use high magnification; all that is achieved is greater camera shake. A wider field of view will show more stars for comparison. Try to record rooftops or trees at the beginning or end of a satellite pass for scale.

As satellites move across the entire sky, using any of the following devices may be best hand held. With practice, a steady image can be obtained. However, it does take time. If a satellite is heading west to east via the South, then keep your feet facing south and rotate your body to the West to begin recording the satellite. Turn your body with the satellite but keep your feet still.

In other words, face the centre of the track before the recording starts and turn your upper body with camera in hand to the satellite. Again, practice on recording aircraft during the day.

Digital Single Lens Reflex Cameras (DSLR)
Many of these models have video recording facilities with sound. Set the sensitivity to at least ISO 1600 setting first. Use a Video compatible memory card, this ensures rapid recording. Also observe the MB/s transfer rate, 45MB/s should be the minimum; the figures are marked on the card. The camera battery will need to be fully charged too, such recordings take a high toll on battery power.

Camcorders
If the camera has Auto-focus only, then it is not suitable. It will rarely be able to keep constant focus on dots. The lens will just be stuck in recording a blur. I am very suspicious of all these UFO reports taken by a camcorder with auto focus. The points of light always seem to become saucer shaped when the filmmaker zooms in. I get exactly the same effect if I zoom in on Jupiter, or the Space Station etc. I know exactly what I am seeing, but the camcorder will show an apparent flying saucer instead; the multiple elements in the lens system cause the effect. Inexperienced observers swear they have recorded aliens. Manual focus on infinity is the only way round this.

Some camcorders are ideal for recording the brighter satellite passes. If you are to purchase one specifically for this, then ensure it has zero Lux sensitivity, manual focus and HD quality. If your camera uses memory cards, then only use Video compatible cards and the minimum transfer rate should be at least 45MB/s; I use 70MB/s.

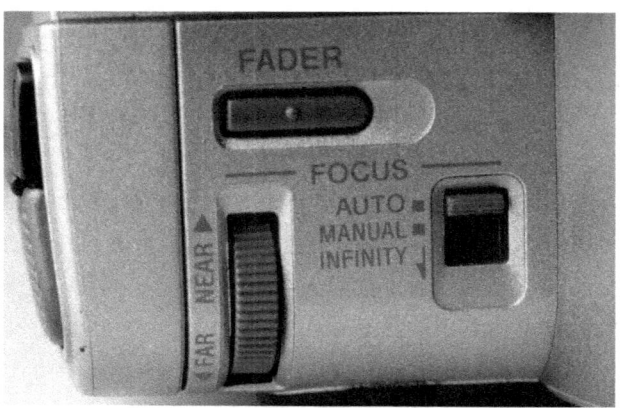

If you are very patient, set up your camcorder on a tripod and zoom in on the Moon when it is full or a couple days either side. Set the focus on infinity and just record. Some auto-focus cameras will be fine for the moon. Every so often something will fly in front of it; birds, planes and yes satellites. If a phase is chosen say 3-4 days before full moon, it is possible to view it during daylight hours of late afternoon and record satellites transiting it. The same can be said for the Sun at any time during the day; if its projected into a dark box, then satellites will be silhouetted against the solar image.

Play recordings back on a TV and wait. There may nothing on it for ages. As each satellite is observed, then that part of the recording can be edited and put aside to add others to it.

Check out our website www.outerspacebooks.com to see such recordings via YouTube. If viewing direct

YouTube, please do ignore any comments regarding 'flying saucers' or 'fake satellites' - they are indeed satellites.

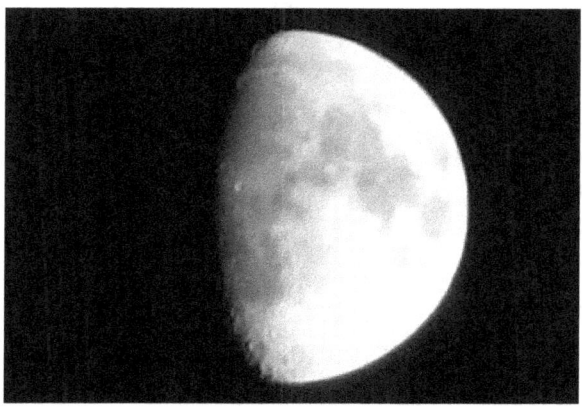

This image was achieved on a HD camcorder with auto-focus. There is a lot of detail on the Moon and very bright for focusing, but recording faint points of light is unrecognisable for auto-focus systems. Seek out manual focus cameras for recording satellites.

Night Vision & Binocular Cameras

The camera shown below was designed and advertised as a day & night wildlife-recording instrument. There are two objective lenses on the front, but only one eyepiece at the back. A CMOS chip that records high-resolution video in PAL (UK) format takes the other up; US NTSC version is available. I split the signal with an adapter; one to a monitor, the other to a video deck or DVD recorder.

At night, this instrument becomes incredibly sensitive to light. It can record stars and satellites down to around Magnitude 9, well below naked eye visibility and far beyond any standard camcorder. Most of my own clips on the website have been recorded using this system.

I built the unit above while self-isolating from the Coronavirus. The whole unit is mobile and can be taken anywhere in the garden. The camera is permanently attached; I just turn on the camera and alter recorder to use Line3; the phono-socket. I press record via the remote when ready with the satellite in view. I can commence recording satellites passing overhead that are hundreds of miles up within 45 seconds of set-up. I do not live long enough to waste time faffing around.

The tripod shown is a video dedicated model. The two main axis for vertical and horizontal is hydraulically dampened to reduce camera shake.

The camera is by Yukon – 5x42mm. The monitor shows a live image from the binocular/camera. This is the best way of ensuring accurate focussing and that all the cables are connected. I have spent many hours recording the night sky only to discover a video connector became detached on the ground or was simply out of focus. A monitor reduces errors. I use a video deck instead of DVD recorder simply because I can obtain them for around £10 ($13) now on eBay. In addition, a DVD needs to be finalised before playing on other devices. A new disk has to be inserted every time. A tape does not have this problem. I super scrimp everything; no point in splashing out.

If the number of 'lines' on the camera is the same or less than the recorder, and then no quality will be lost. In the previous example, the wildlife camera produced 400 lines and the Sharp video recorder – 600 lines. So no information is lost as compared to an expensive HD DVD recorder of over 1000 lines; no point if only 400 lines are produced anyway.

To focus this type of camera, two stages are required. First focus on a bright star and observe your monitor to ensure you have the correct set-up. Then refine the focus onto fainter stars as the brighter objects tend overload the recording chip and become a blob. The fainter stars remain as small points so they are easier to home-in on. You are now set for satellites.

My Yucon wildlife camera is perfect for recording stars and satellites. It is sensitive enough to record all objects within reach of a standard pair of 7x50 binoculars.

Models of binoculars are now available that have built in digital video cameras. They include a screen and an easy-to-use menu. The objective lenses are not that large though; I am not too sure as to how faint a satellite or star these can record.

As these instruments are not driven automatically on a steady telescope mount, hand held (with practice) or at least video-tripod based techniques are best. It is possible that they could be 'piggy-backed on a driven telescope. Search eBay for 'Binocular Camera.' They have manual focus. They use SD memory cards; again do use dedicated video SD cards.

Do ensure such models are Image Intensifiers rather than just night vision. The latter will not work on the night sky very well.

If money is no object, there are rare binocular models that have detachable eyepieces. One can be used to piggyback a camera instead of looking through. The other eyepiece you will still be able to use for tracking and focusing. These will not be driven to follow satellites or even the stars, but will certainly be enable you to record satellites in real time out to tens of thousands of km. The next pair shown is by Vixen and can be found new or used on eBay for around £2000 ($2500) used as a guide.

These eyepieces are removable for possible camera attachment; visit your local camera store or Astronomy club for advice.

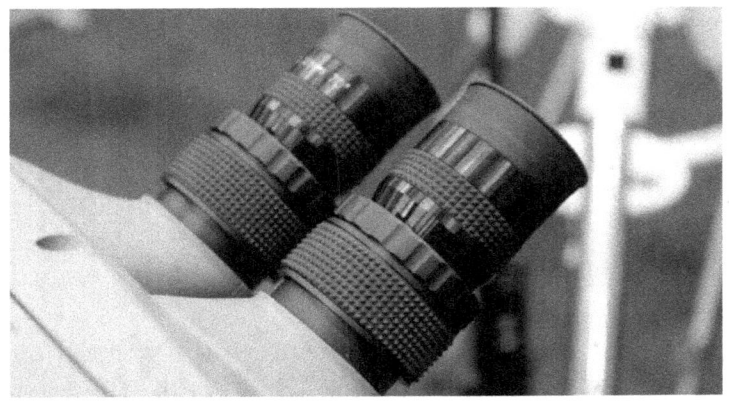

All Sky Cameras

These come in various forms. The have a very wide-angle lens from 140° to 170° so they are capable of covering virtually the whole sky. Choose the highest definition camera you can afford and the most sensitive. They can record all aircraft passing over, as well as meteors. They can give out a standard PAL or NTSC video signal. Some are designed to record on a laptop. They can be as low as £75 ($100). Fish eye lenses for digital SLRs are perfect for this too but can burn a hole in your pocket. Do investigate eBay for these and get the best deal possible.

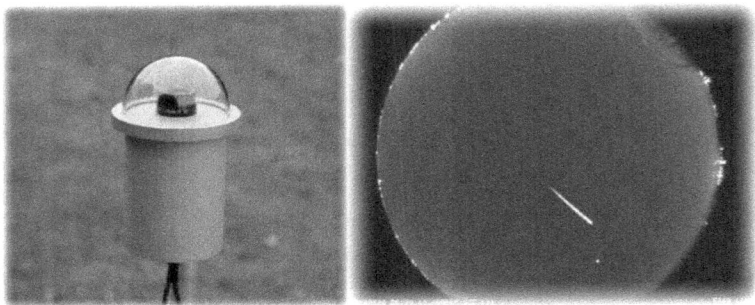

A meteor captured with an all-sky camera from Huntsville, Alabama.

Chapter 26 Telescopic Imaging

Photographing moving objects through a telescope is incredibly hard. A good telescope has a drive motor to follow the stars. Satellites move in a different direction and speed compared to the motion of the stars due to the earth's spin.

Satellites that are very high up, around 10,000 km or more may have a chance of being imaged as they move slowly enough in your field of view to capture them. The example below is of Apollo 8 when it was around 50,000 km from earth.

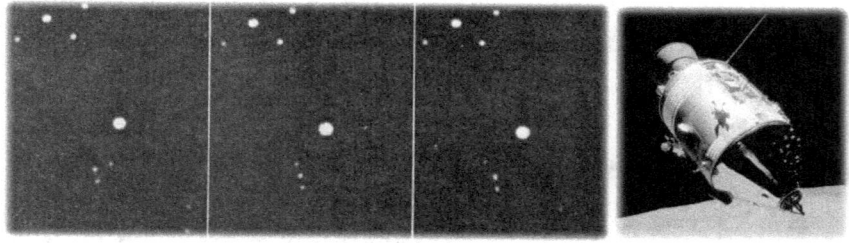

Above; The Apollo 8 Spacecraft on its way to the moon, 22nd Dec 1968. Photographed by the Yerkes Observatory, Chicago.

Geo-stationary satellites are just that; stationary above the earth's equator. They remain in the sky as they are rotating at the same speed the Earth is; once every 24hrs. The telescope can now remain perfectly still, no drive working. Use the lowest magnification for a wider field of view to increase the chance of imaging the target and reduce amplified vibration from wind etc. Set it up on the general satellites' position discovered using Heavens-Above and take a 10-second exposure or so.

The result should be trailed stars and a satellite or two as a point.

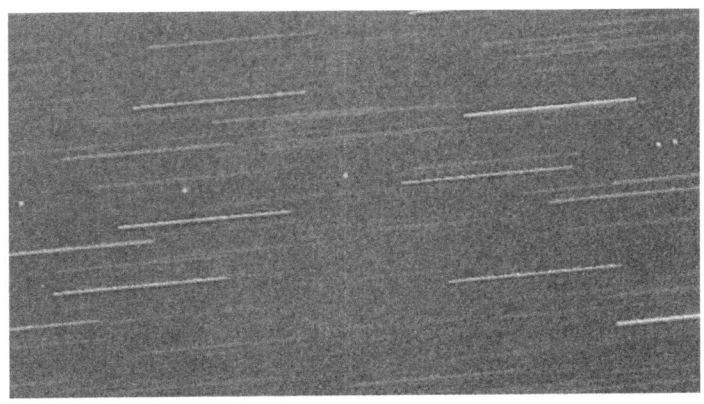

Above; five geo-stationary satellites imaged in 2009 through a low power telescope. These are positioned around the Celestial Equator in the sky. Check a star map for your calendar month.

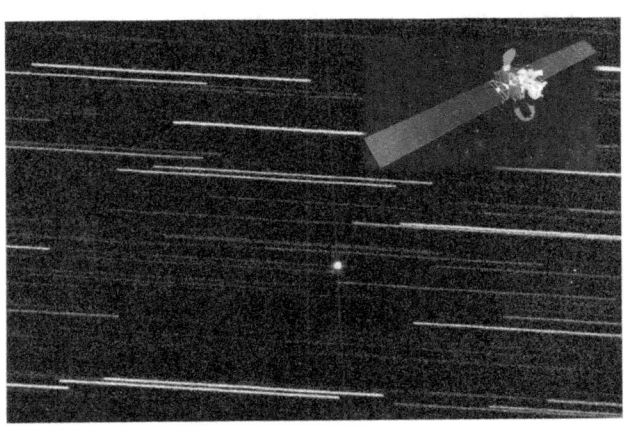

Image of Geo-stationary satellite Intelsat 706; taken through my 5" refractor. It was given a 20 second exposure; the stars trailed across as the Earth rotated, the satellite was a giveaway as it orbits; matching the earth's speed.

It you have a programmable drive system on the telescope that can slew rapidly across the night sky, it is possible to take close up images of the larger satellites that are less than say 1000 km up.

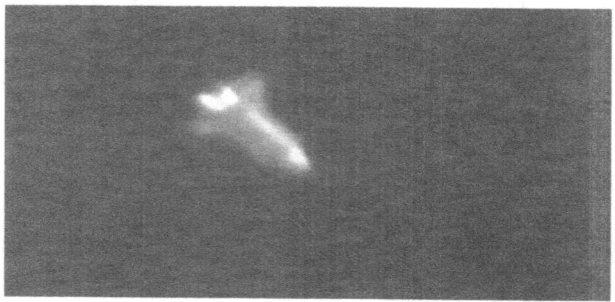

The Space Shuttle Endeavour on STS 57 Mission 23 June 1993.

Above; Envisat photographed by Josef Huber of Germany in 2005. The image of Envisat is on the left compared to a model in the same orientation to the right.

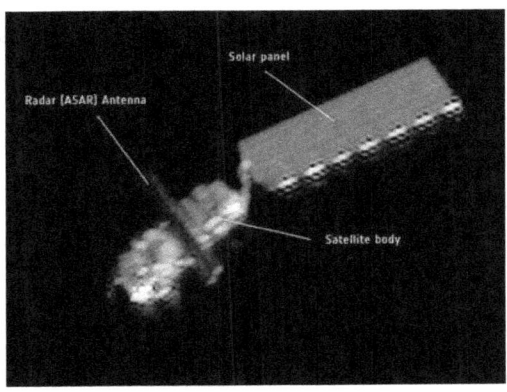

A picture of Envisat taken via another satellite in orbit 2014. Such images are often classified. Credit ESA.

Anyone who wants to investigate driving a telescope to follow a satellite can explore websites such as www.heavenscape.com.

Not all telescopes are compatible. Some Celestron NexStar telescope range are and the Mead Autostar LX200 telescope range certainly is.

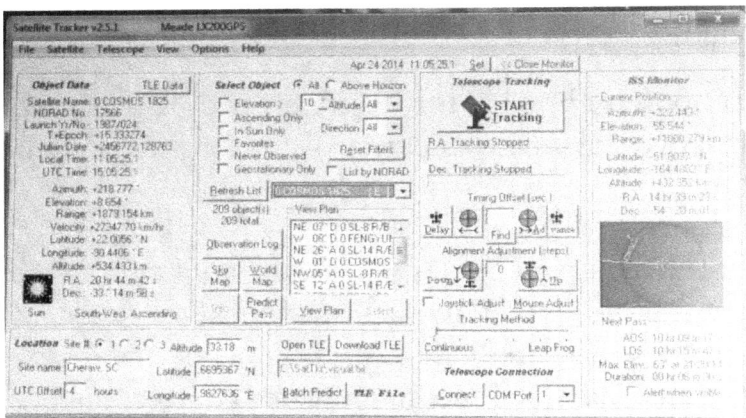

(More about this process will be published on the website in the near future. Obviously, a printed book cannot keep up-to-date in the same way).

Chapter 27   Simple Image Processing

Once you have an image of a satellite passing over, you may be a little disappointed with the picture itself. Perhaps there are not as many stars as you would like or light pollution has caused a red or green glow across the whole image. With film photography, there was very little than could be done to improve it, you were stuck with what you had. With digital, the image can be rescued to a high degree. (A white glow from LED street lights cannot be altered as stars are generally white too).

Free versions of advanced programs are sometimes available such as Photoshop PS2 but rare. I use an old program called Photoshop 7. It can be purchased on a CD ROM on eBay for around £5 ($7.00). I find this more than adequate for touching up starry photos. Other free or low-priced image processing packages are around.

This book is aimed at the beginner; people that perhaps have never taken any pictures of the stars and satellites before.

There are some processes that can be applied to scanned images of film that reduce scratches or grain; these should never be applied on starry images. The software will think the stars and satellite trails are scratches or dead pixels etc and so 'intelligently' remove your entire image. Therefore, there are only a small number of key features that need to be learned to improve your satellite trail images.

Example

I have chosen one image to improve, although it does not include a satellite. The original picture does not show many stars and light pollution has altered the true colour of the sky. You may recognise the constellation Orion and Jupiter is seen to the right.

Step 1 is to change the overall colour of this yucky green. Use 'colour balance' and play with the slider controls until you find a more pleasing and natural colour. You can try clicking on 'Auto Colour' instead but it can go too far. If you do not like the result, just click 'undo' or 'cancel.'

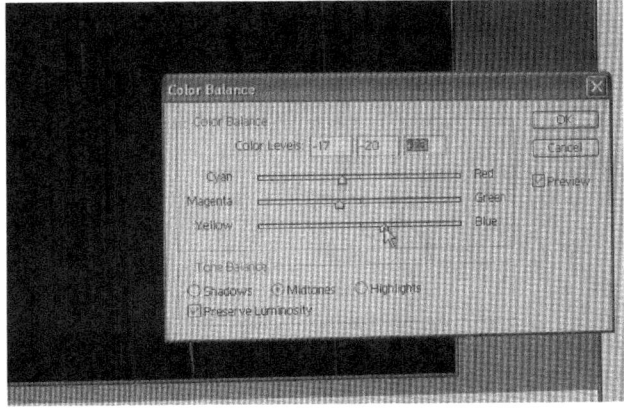

The next step is to improve the number of stars in the image. The CMOS chip almost certainly captured some data of stars that do not show on your image. Using a facility called 'Curves' you can brighten the star data and keep the background sky dark at the same time. As soon as you click on it, you will be showing a graph with a straight line. Using graph co-ordinates go along three and up three. Click on the line and move it up roughly to the point shown. The stars will now brighten and more will appear. The downside is the sky will brighten too. The next step will cure it.

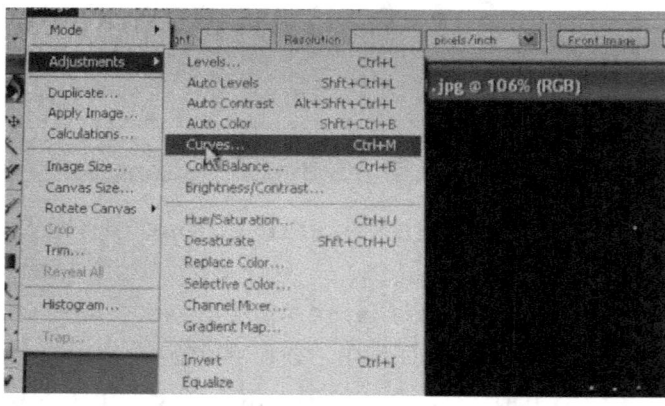

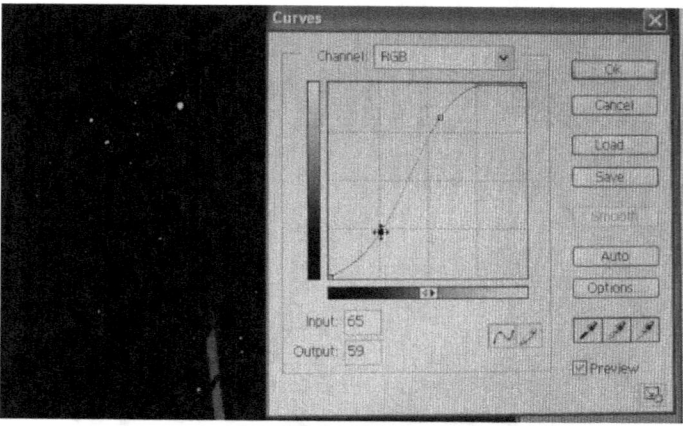

The rest of the line is now a curve, grab the line at around one across and two up and drag it down roughly to the point shown. This will now darken the background sky without dimming the stars. The best results will be from experimenting.

Simply study the graph; the two sides show dark to light. Move the dark part of the straight line that is shown to you a little toward the bottom to make that part of the picture darker still, and the lighter part of the line upward to brighten the stars further. More information that was hidden is now shown instead and the background sky darkens to reduce light pollution.

Save the picture with a new name and keep the original. As time passes, you will learn extra skills and may wish to alter the original image differently at a later stage.

The green glow has now gone and the number of stars has trebled. A satellite trail may have been recorded perfectly but seem faint on your original image. These two techniques alone will save your picture and motivate you a little more perhaps in achieving even better results in the future.

Try using the contrast option too but not too high a setting. A faint satellite image may be best displayed if the image was turned into a Negative or even Black & White.

One more example of before / after comparison...ISS is seen fading into the Earth's shadow. Photo by a long-term friend; Tony Rickwood of Ullapool, Scotland.

Chapter 28 Favourite Targets

There are numerous targets in the sky for beginners as well as experienced observers. I have chosen a small number that are interesting from a human standpoint, observational interest and historical.

ISS

By far the brightest and most well-known satellite of them all is the International Space Station. When this fantastic piece of engineering passes over your hometown, it really brings home that this technology is for all of us to wonder at and be proud of. It should remain operational and seen in orbit until at least 2025. The Sun is the brightest object in our skies; the Moon is next; this is in third place.

NASA image; Spotting and imaging this satellite is an excellent way for anyone get interested in this hobby. On board will be human beings flying overhead at 7km per second. It can be seen twice an evening or morning.
Do not forget to wave as they go by.

You may also be lucky enough to witness a docking or undocking event. A second fainter object will then be seen a few degrees from it.

As of November 2014, Russia is currently looking at pulling out of the ISS project in 2020 and building their own space station once again. A political decision rather than a scientific one regarding the troubles in Ukraine and the US / European / Russian relationship probably cause this initiative.

ISS is by far the easiest satellite to image. This is a 10-second exposure; Jupiter is the bright star to the lower left of the trail. The tree provides scale.

Mars Ship?

It is possible to transform ISS into a ship capable of ferrying a crew to Mars. It has all the facilities required – power, crew quarters, observation deck, food storage etc. The only extra hardware required would be two Mars decent vehicles to be docked on and a re-startable engine. Its current velocity is already 70% that is required for a journey to Mars (My idea – several astronauts have agreed to it in principle including John Blaha – visited ISS several times via Shuttle and Soyuz).

Envisat

Envisat was an Earth observation satellite. Its objective was to provide additional observational data to improve environmental studies.

In working towards the objectives of the mission, numerous scientific disciplines use the data from the different sensors on the satellite. They study atmospheric chemistry, ozone depletion, biology in the Oceans, sea temperatures, wind strength, waves, humidity, floods, agriculture natural hazards, digital mapping, sea traffic, atmospheric pollution, snow & ice.

Envisat cost £2billion including operational, development and launch costs. The mission is now replaced by the Sentinel series of satellites, but Envisat is still in orbit for us all to see.

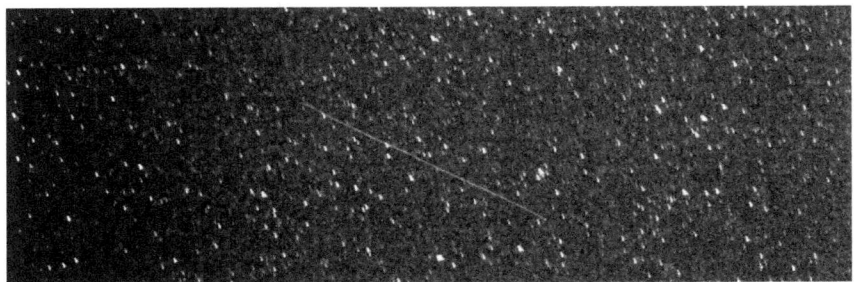

Envisat Trail over Huntsville, Alabama.

Size of target 26mt (85ft) × 10mt (33ft) × 5mt (16ft). Its size and low orbit defines a constant brightness of between Mag 2.5 & Mag 3.5; bright enough to be seen from most towns.

Its orbit is 100 minutes long, 774 km high almost circular at 98° and repeats the ground track every 35 days. Therefore, this is visible most of the year from all over the world. In 2012, contact was lost and is drifting in orbit, but it will remain visible for decades to come until its orbit naturally decays.

ESA image.

Tiangong 2
China's first space station was Tiangong 1; serving as both a manned laboratory and an experimental test bed to demonstrate orbital rendezvous and docking capabilities. It was launched unmanned aboard a Long March rocket on 29 September 2011. This space station ended its life in April 2018 and was replaced by Tiangong 2.

This is a very bright satellite, often passing over the southern parts of the UK but low down, as the inclination is only 42.8°; the south coast of England is 51°. It can only reach a maximum of 20° above the southern horizon.

Explorer 7
Launched on 13 October 1959 by a Juno II rocket from what was called then Cape Canaveral into an orbit of 573 km by 1073 km and inclination of 50°. It was designed to measure solar x-rays and other energetic particles and heavy primary cosmic rays. Secondary objectives included collecting data on micrometeoroid penetration and molecular sputtering and studying the earth-atmosphere heat balance.

The satellite mass is 41.5kg, 75cm x 75cm. Powered by solar cells it also carried 15 nickel cadmium batteries around its middle to even out the mass to stabilise any rotation. It transmitted data continuously through to February 1961 and went dead 24 August 1961. The orbital inclination is at 50° therefore visible out to 55° or so north and south of the equator. This includes the UK and is one of the oldest satellites that can be seen.

However, due to its small size, it is not very bright, maximum of Mag 5.5, only just visible to the naked eye in perfect conditions. Easy binocular object when it is at lowest point of its orbit – now at 520 km due to atmospheric drag. Search Heavens-Above for its position.

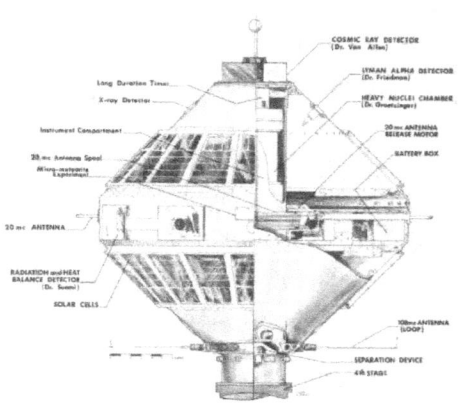 

Prospero
The only satellite built and launched by the UK. The object is only around 1mt across and its minimum altitude is 527 km. This is only visible in binoculars and telescopes, but is very rewarding for folk in the UK. This was launched in 1971 on a Black Arrow rocket at Woomera, Australia. Black Arrow rocket at Woomera, Australia. The satellite is no longer transmitting.

A full size model of Prospero at Alum Bay on the Isle of Wight. Walk to the top of the hill facing the Needles Lighthouse and pass round the corner. A free museum houses many artefacts from this era.

A full-size replica of the Black Arrow is on display at Sandown Airport, Isle of Wight.

Then and now 1969 / 2018 from a drone. The UK launch test facility at Alum Bay, Isle of Wight UK is open to visitors and completely free. Lower photo by the author.

Prospero - Visible Passes

Search period start: 17 May 2014 00:00
Search period end: 27 May 2014 00:00
Orbit: 527 x 1313 km, 82.0° (Epoch: 16 May)

Passes to include: ⦿ visible only ◯ all

Click on the date to get a star chart and other pass details.

Date	Brightness (mag)	Start Time	Alt.	Az.	Highest point Time	Alt.	Az.	End Time	Alt.	Az.	Pass type
17 May	8.4	22:06:12	10°	SW	22:12:24	32°	WNW	22:17:41	10°	N	visible
18 May	8.8	22:15:53	10°	WSW	22:21:25	25°	WNW	22:26:13	10°	N	visible
19 May	9.1	22:25:47	10°	WSW	22:30:28	20°	WNW	22:34:37	10°	NNW	visible
20 May	9.3	22:35:58	10°	W	22:39:33	15°	WNW	22:42:50	10°	NNW	visible
21 May	9.5	22:46:44	10°	WNW	22:48:41	11°	NW	22:50:31	10°	NW	visible
25 May	9.2	21:34:54	10°	WSW	21:39:06	19°	WNW	21:42:48	10°	NNW	visible
26 May	9.3	21:45:03	10°	W	21:48:07	14°	WNW	21:50:55	10°	NNW	visible

Use the 'Search Database' button on Heavens Above website to look for a specific satellite. As I became interested in satellite spotting, this was my first major ambition; to see and record Prospero passing over Britain.

Check out the 'British Space program' page on www.outerspacebooks.com for a clip of it passing over the UK by the author.

The first satellite on this clip is Cosmos 1244 and Prospero is seen half way through the clip as the fainter object travelling in the opposite direction. The e-book version of this publication will take you straight to such clips.

CryoSat 2

CryoSat 1 was a launch failure and fell into the ocean. But CryoSat 2 was successfully launched in 2010 and partly built in the UK within Astrium. Duncan Wingham of University College, London, made the original proposal for the project. For this mission that includes 3D mapping of ice, extremely accurate satellite positions are required. This is achieved by a laser retroreflector on board and a radio DORIS system that works by radio parallax from the ground.

After three years of construction at Stevenage, Hertfordshire, final testing took place in Germany and launched from Kazakhstan. It is under contract from the European Space Agency to monitor the changing conditions at the earth's poles. The orbit you may have guessed is polar and passes over Britain every day. Its average orbital height is 710 km and is only 4mt x 2.3mt in size on the downward facing axis. This object is not very bright; around Mag 4, but is largely a British built satellite and gathers environmental data; the ultimate purpose of 21st Century space technology.

SeaSat 1

Launched in 1978 from Vandenberg Air Force Base, California. It is in an almost circular polar orbit around 745 km high. This is a relatively slow crossing satellite; about 70% of the speed that ISS takes to cross. Its brightness is unusually high due to the solar panel surface area at between Mag 1.9 and 3. So the combined points make it a relatively easy target to photograph or capture as a moving image from any country.

Spy Satellites

Early spy satellites such as the US Corona series used to take photographs on black and white or Infra-red film Then it released a capsule with the exposed film on board These *'Buckets'* were picked up by a military team. They were designed to be retrieved during the parachute stage of recovery. They could float, survive desert or artic conditions, and transmit a location beacon on a secret frequency. The movie *'Ice Station Zebra'* is weakly based on a true story. The USA & the Soviet Union both launched such machines.

Modern spy satellites use digital cameras of course and have superseded the Buckets. They can now have an operational lifespan of a decade or so rather than weeks. The easiest to find is the US Lacrosse series; relatively bright and in a high inclination so most of the world can view them and most of the world is monitored from them too.

The cameras are always being superseded by new technologies developed down here so such satellites are upgraded often.

They can also fly in formation in twos or threes as a triangle around 10 degrees apart.

One story that has leaked out regarding an early Lacrosse satellite goes back to the 1990's. Colonel Gadaffi was hiding as usual somewhere in the Libyan Desert. The CIA wanted to determine his exact position and employed the best spy satellite they had. Gadaffi employed a British (sorry about that) satellite spotting expert to predict when each US spy camera passed over his camp and blew a whistle minutes before each pass. Everybody dived into their tents.

Gadaffi became tired of this game and purchased some Colonel Gadaffi masks. They were given out to everyone on the camp and next the next time the whistle blew, everyone donned the masks, stood still outside and faced the sky. Gadaffi himself wore a mask as an infrared image combined with optical may have shown whom the real one was. The CIA realised that they had been rumbled and gave up.

Hubble Space Telescope

The orbit of the HST is at an inclination of 28.5°. Taking into account the angle of sight from Earth, it can only be witnessed between around 38° north or south of the equator. Southern UK is at 51°; no chance here. The reason for this placement was simply economy. The closer the orbit is to the equator, the more advantage can be taken of the earth's rotational spin; less fuel. The closest equatorial orbit achievable from the Kennedy Space Center is 28.5°. Observers in the southern US will see it.

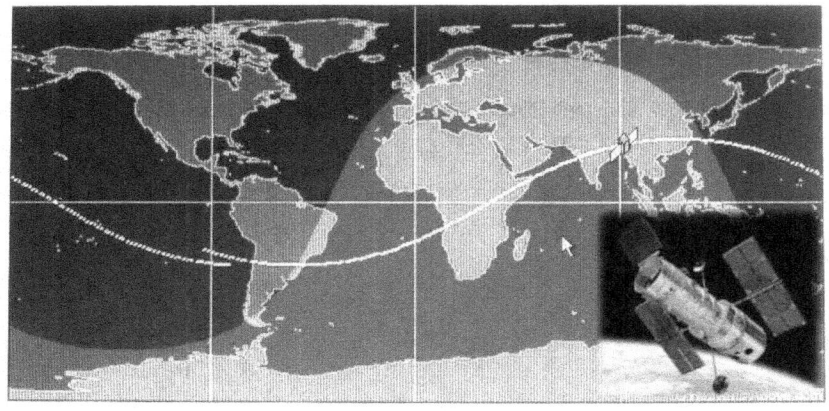

(This has an extra special meaning to me as I saw it launched by the Discovery space shuttle in April 1990).

The satellite prediction maps include the whole sky as if viewed by a fish-eye lens. The satellite path will be shown as a curve.

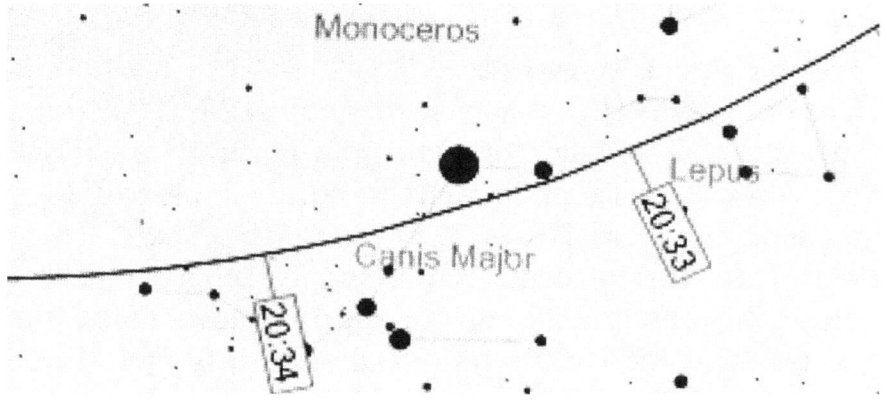

The exposure below is of the Hubble Space Telescope from Huntsville, Alabama in 2014. It was just one degree lower than the predicted path. This can be a bright target due to its size and the orientation of its solar panels.

The bright star is Sirius in Canis Major to the lower right of Orion. You can make out the whole of Lepus the Hare to the right of the image.

ICESAT 2

This satellite launched on 15th September 2018 from Vandenberg Air Force Base, California on a powerful Delta II rocket. Its function is to map accurately the thickness of sea and land ice at the poles, glaciers and mountain ranges to within 4mm of accuracy; a measurement totally impractical and financially unfeasible to gain from the ground. A complex on-board system of laser-based radar (Lidar) allows this data to be accumulated and build up a global picture of ice trends connected with climate change.

This is a bright satellite that can be seen the world over due to its polar orbit. The solar panel can produce flares that can brighten its appearance from Magnitude +3 to -1. Search Heavens-Above for timings from your local area. It is well worth a look.

This trail of ICESAT 2 near the Lagoon Nebula was taken from Arizona in April 2019 by the author.

Starlink;
A constellation of satellites constructed by the SpaceX Company to provide low-cost internet to the globe. When complete, it will consist of thousands of mass-produced small satellites working with ground transceivers. SpaceX also plans to sell some of the satellites for military scientific or exploratory purposes.

Starlink constellation phase 1 will consist of 1,584 satellites at 550 km altitude. As of January 2020, SpaceX has deployed 182 satellites. They plan 60 per launch, as often as every two weeks after late 2019. In total, nearly 12,000 satellites will be deployed by the mid-2020s, with a possible extension to 42,000. Commercial operation begins in 2020.

Concerns have been raised about the long-term danger of space junk resulting from placing thousands of satellites in orbits above 1,000 kilometres (620 mi) and a possible impact on astronomy, although SpaceX is reportedly attempting to solve the issue.

SpaceX estimated the total cost of the decade-long project to design, build and deploy the constellation in May 2018 to be about US $10 billion; the project began with the first two prototype satellites launched in February 2018. A second set of test satellites and the first large deployment of a piece of the constellation occurred on 24 May 2019 when the first 60 satellites were launched. A dedicated page is on the website.

These satellites cross the sky in a line. Up to five can be seen at once, with dozens more following. From 2020, these are probably going to become the most favoured of all the satellites to be seen in the sky.

Chapter 29   Transiting Satellites

Every now and then, satellites, as well as aircraft and balloons can pass in front of the Sun and moon. This is known as a Transit. The planets Mercury and Venus can Transit the Sun on rare occasions. As the Sun and Moon are just 0.5° across and most satellites travel at around 0.25° per second across the sky, photographing the event is somewhat tricky; but not impossible. The higher satellites in mid-Earth orbit or above take several more seconds to cross.

Several satellites a day will Transit the Sun or Moon and can be imaged even without using the Internet. All you need is one skill – patience. The best time of the year is when the Sun crosses the equatorial plane – 21 March & 21 September. This is when the geo-stationary satellites can transit the Sun as they are directly above the equator; so will the Sun.

When I was about 13-14yrs old, I imaged the Sun every single clear day when I was off school. I waited till mid / late afternoon when the Sun shone into my bedroom, pointed my telescope toward the Sun and projected the image onto a large white card. I could produce it up to 2ft across. I darkened the room with thick blankets but allowing the sunlight to travel into the telescope. I observed the sunspots change shape by the hour. Every now and then a plane would fly in front of the Sun and saw its silhouette complete with vapour trail.

On rare occasions, a small dot would take a few seconds to cross the Sun and I was baffled as to what it was.

'UFO' I thought straight away. After a little research I realised I was seeing satellites. 'Ohhh shucks.'

My solar projection method in 1977. A thick blanket was pegged up to darken my bedroom and the telescope was pointed out the window toward the Sun. Tracking the moving Sun was made by turning the telescope controls.

With experience using satellite prediction websites, preparation can be made to see the larger satellites transit the Sun or Moon. However, please never look through any instrument at the Sun unless you are using the appropriate filter please.

The space shuttle Atlantis transiting the Sun on mission STS125 in 2009. Photo by Theirry Legault from Florida.

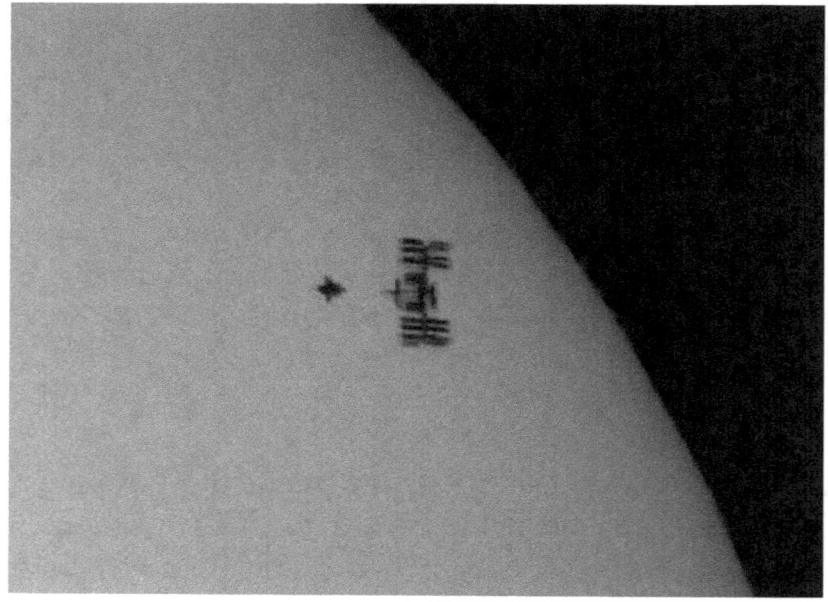

Another incredible shot by Theirry Legault of ISS and the space shuttle Atlantis in 2010 from Spain this time.

Orbiting satellites occasionally pass in front or very close to the Sun in the sky. The Sun is very noisy in the radio spectrum and often interferes with the satellite transmissions. These 'Sun-out' events with Microwave transmissions hardly suffer from this effect, as the Sun is quiet in the microwave frequencies. All the TV satellites use microwaves for this very reason.

An image of ISS transiting the moon.

Ed Morana specialises in satellite transits. Try his internet page for other links on how to observe them yourself. Such information changes daily so this book cannot keep up. The e-book makes these links more accessible / or use the website and look up the page on 'Transiting Satellites.'

Chapter 30   Re-entry

Even though satellites are generally orbiting in the vacuum of space, there is still a trace of atmosphere buffeting the machine. There may be just a few air molecules a second hitting the satellite that may have a mass of several tons. This weak force can have an effect on the orbit; month by month it will lose speed and drop gradually toward the earth. This in turn will increase the number of air molecules encountered per second and accelerate the process.

During a satellite's useful life, on board engines will fire to boost its altitude back up toward its parking orbit. After ten years or so, either the fuel will run out or the satellite has now served its purpose. The orbit can now be allowed to decay and finally enter the earth's atmosphere. The final orbit will only be around 65-70 km above the earth.

The very first confirmed crash landing of a satellite was a piece of Sputnik 4 that landed in Manitowoc, Wisconsin, USA in 1958. It was sent back to Russia and never received any thanks. A metal ring in the street marks the historic impact point to this day.

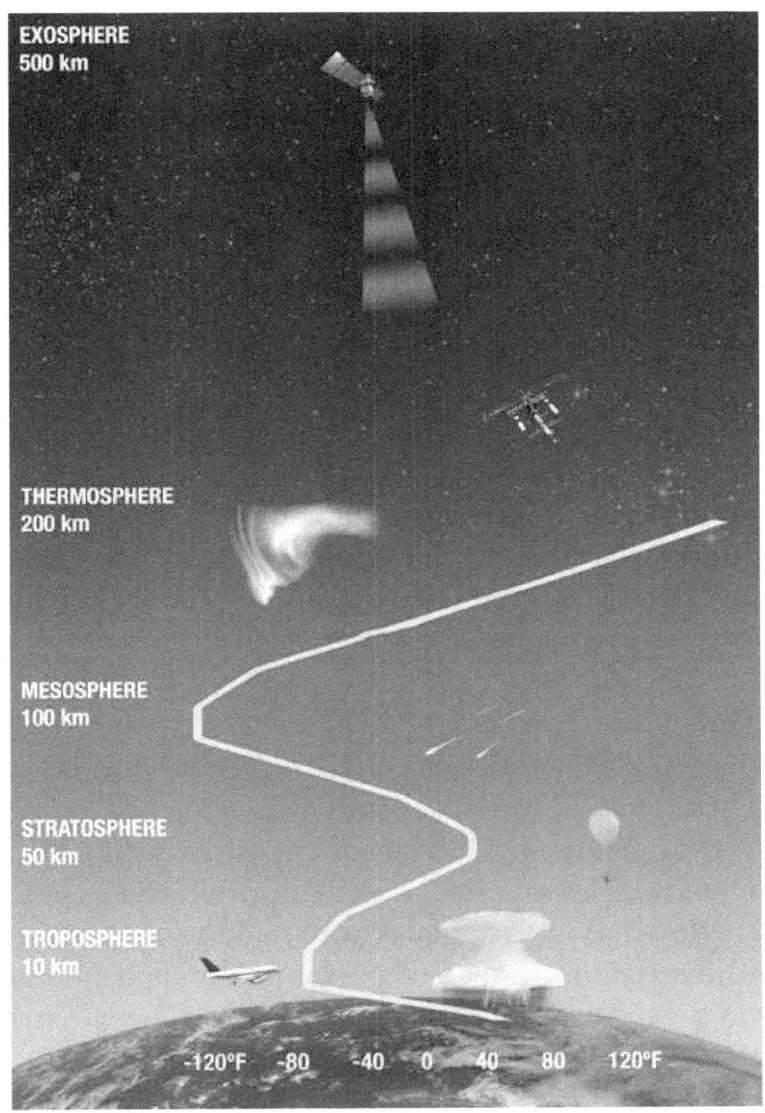

Even though satellites are hundreds of kilometres up well into the Exosphere, these still is a few molecules of air per cubic centimetre. This is enough to slow a satellite down over time. This will reduce its velocity and bring it lower still, more are molecules are encountered per second so slowing it further.

Satellite re-entries can be quite spectacular and viewed by large numbers of people as the craft will often be seen over a few thousand km before it finally succumbs. The first example was Sputnik 2 and was witnessed by many people during its descent on 14 April 1958. For ten minutes it travelled from New York to the Amazon, it descended from 130 km to around 60 km, leaving a trail of `sparks' some 100 km long, the satellite itself produced a multitude of brilliant colours.

To reduce the increasing hazard of defunct satellites, a global policy now exists to reserve some remaining fuel to deliberately slow the satellite down at this last stage and force it to burn up within a few days. Such events are a little more predictable than the random re-entry. This is known as the de-orbit procedure.

Skylab

Perhaps the most infamous of all the re-entry events was Skylab. This 77-ton spacecraft was launched in 1973 and was operated by 3 teams of US astronauts. After its last mission, it was left to drift in orbit and began to come down. By 1975, plans were underway for a space shuttle mission to attach a booster to it and push it back up to a safe parking orbit. It would have been carried out on the shuttle's fifth mission sometime in 1980. As time passed it became obvious the shuttle wouldn't be ready in time, the first launch date slipped further toward 1981.

The shuttle program was plagued with technical delays and the Sun became more active. This expanded the upper atmosphere and increased drag on all low satellites A new predicted re-entry date was sometime around May 1979. This was a result from a study carried by satellite spotters (myself included) timed and plotted the orbit and submitted them.

The North American Air Defence Command (NORAD) located the space station by radar, aimed a radio signal at it and received an echo. For two minutes, Skylab transmitted on the condition of its systems. It was rotating at about 10 times per hour and when its solar panels turned sideways to the sunlight the radio transmissions ceased as there was no electricity. This spin was confirmed by amateur satellite observers by measuring the changing brightness during each pass.

The first thing the engineers needed to do was to charge the batteries and since they could transmit commands only briefly once during each orbital pass, this would take time. Within a week, they had charged two batteries. The onboard computer was used to help control the spacecraft via its Reaction Wheels. The aim was to decrease drag by rotating the panels parallel to the Earth and keep it in orbit longer for a shuttle mission, but too much altitude was lost already.

During its last moments, NORAD managed to rotate the spacecraft in a sideways orientation to increase drag and force it to comedown somewhere over the Pacific Ocean. As such an experiment had never been tried before and data on the upper atmosphere at that time was limited, it was incredibly difficult to control a re-entry point. It was hoped for it to come down over the Pacific and most of it did. Shortly before 1pm, NASA at their Washington HQ received word that the area southeast of Perth, Australia, had indeed been showered with some pieces.

Spectacular visual effects were reported and many people heard sonic booms and whirring noises as the chunks passed overhead. The majority did indeed come down over the Pacific. This was almost certainly down to good luck on the day rather than planning. Some parts have been moved to Huntsville, Alabama and are on display at the Marshall Spaceflight Center.

A fuel tank on display at the Marshall Spaceflight Center, Huntsville, Alabama. (Image by the author 2010)

Some vehicles are planned to re-enter are under full control by a space agency or company. Such events are far more predictable and remarkable sightings are made. Unfortunately, for many of us they are normally made over an ocean. The next image is of a service vehicle for ISS called the Jules Verne ATV module burning up over the Pacific Ocean.

Satellites that are naturally falling from orbit are announced on satellite prediction websites such as Heavens-Above. Announcements are made to observers to look out for such events if you happen to be under a predicted fall. I have only ever witnessed one re-entry event from a stargazing camp in Sussex, England. I do not know what satellite it was but it lit the sky for around 30 seconds and produced beautiful green and blue colours. These are produced by the different melting materials during break-up.

Several re-entry clips are shown on our website, including one sighting of my own.

Cosmos 954

On 24 January 1978, a fireball streaked across the skies over the Canada. Cosmos 954, a Soviet nuclear powered satellite, crashed near Great Slave Lake, scattering radioactive waste across Alberta and Saskatchewan. There urgent questions for Canada's Prime Minister; *Why wasn't there more warning? Were the Americans and Russians holding back information? And who will clean up the mess? Who is going to pay for it?* The Soviets denied it even existed.

A portion of the craft fell near a hunter's camp. A canoeist travelling through found it, looked at the unusual item and then left it alone. A massive search was begun to locate the rest of it. Hundreds of troops from Canada, Britain and the US joined the search that lasted until October. Tight security was present and no civilians were permitted. The canoeist and his radioactive find were taken back to Yellowknife; the canoeist was found to be in good health.

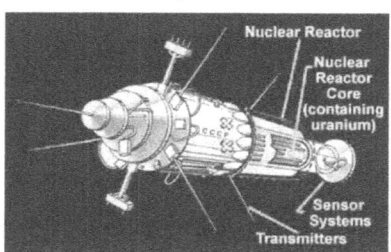

After the clean-up, the Canadian Government sent a $15 million bill to the Soviets. They paid less than half of this amount and agreed not to take back the spacecraft. The Canadians were happy with the amount they received but were happier still that the Soviets had finally admitted the spacecraft's existence.

There was a request from the US to stop satellites containing radioactive material from orbiting the earth. This was followed by similar cries from Canada and Europe. In November 1978, a ban on nuclear-powered satellites in Earth orbit was agreed.

MIR Space Station

Until the construction of the International Space Station, (ISS) MIR was the largest structure ever built in space. The inevitable end was bound to be a fiery one that required careful planning or somebody could get hurt.

The de-orbit of Mir was a controlled atmospheric re-entry of the modular Russian space station. With MIR at the bottom of the ocean, it allowed Russia to concentrate its efforts on the new ISS project.

It was carried out in three stages. The first was waiting for atmospheric drag to decay the orbit an average of 220 km. This began with the docking of Progress M1-5. This was achieved with two burns of the Progress control engines on 23 March 2001. After a two-orbit pause, the third stage of Mir's de-orbit began with the firing of the engines lasting a little over 22 minutes. The re-entry at the altitude of 100 km occurred on 23 March 2001 and broke up in a few minutes.

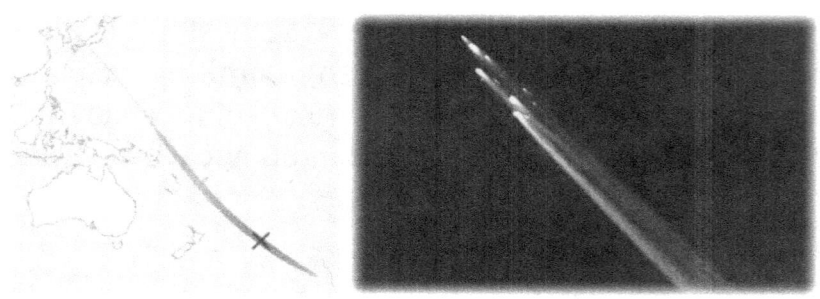

Break up of the MIR Space Station 23 March 2001 over the Pacific Ocean.

Antares Rocket

This map shows the re-entry path of an Antares rocket body that ended its fiery journey in the Indian Ocean as indicated on the yellow track. The blue line represents the predicted re-entry point. As you can gather, it continued for another 40 minutes or so half way around the world. This is not an exact science, so amateur satellite spotters do play an important role in this field.

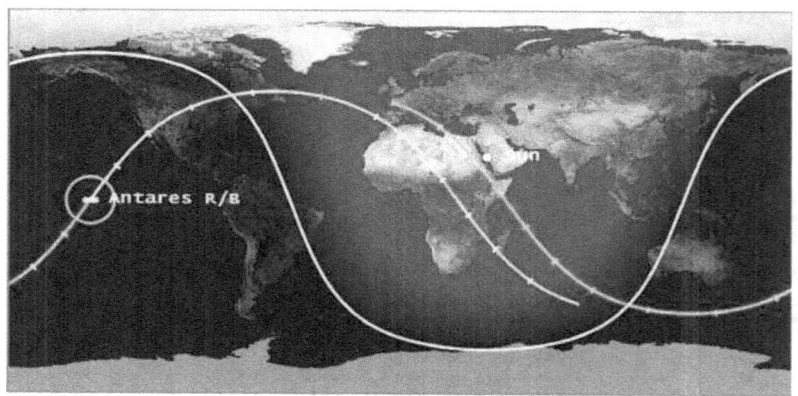

A useful link to re-entry events...
www.aerospace.org/cords/

Point Nemo: An area well away from land has been chosen as a target for many satellites to end up. This is nicknamed as the 'Satellite Graveyard' or Space Cemetery.' Its official tittle is *Point Nemo* - named for the fictional submarine captain in Jules Verne's 20,000 Leagues Under the Sea. This is the furthest point from any landmass on earth, and 4km under the sea.

When their outer space journeys come to an end, old satellites, rocket parts and space stations are aimed to this remote spot in the Pacific Ocean to rest on the dark seabed forever. The technical name for this stretch of water is the *"Ocean point of inaccessibility"* because it lies about 2,700km from any land. The ISS is expected to end up here around the year 2030.

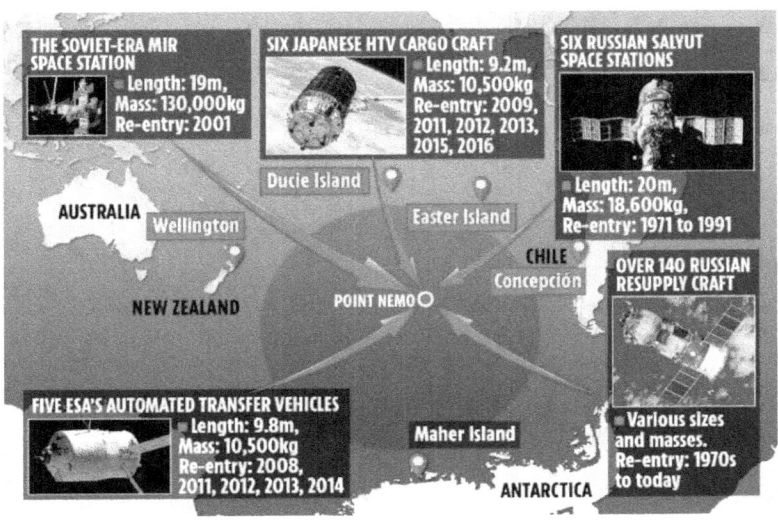

The furthest point from any land mass has been chosen as a target for semi controlled re-entry target for satellites from all nations... Point Nemo.
Credit – The Sun newspaper.

Some vehicles are planned to re-enter are under full control by a space agency or company. Such events are far more predictable and remarkable sightings are made. Unfortunately for many of us they are normally made over an ocean.

On 25th July 2024, my wife and I were visiting friends on Oak grove, California. On our very first night, we witnessed a re-entry event of an unknown satellite. There was no sound, just an incredible display of light with a dense smoke trail behind it. We recorded the sighting on video and took a screenshot below. A full YouTube movie is linked on the website's re-entry' page within the Satellite Spotting section

Satellites that are naturally falling from orbit are announced on satellite prediction websites such as heavens-above.com. Announcements are made to observers to look out for such events if you happen to be under a predicted fall. Different melting materials produce these during break-up. Several re-entry clips are shown on our website.

Chapter 31　War in Space

Our entire lives including jobs, food production, education, finance and business in general is entirely dependent upon communications. In the past, we used the mail system, radio and then we had telephones added to the list. Now we rely on cables and satellites. These are based on high technology either placed deep under water or blasted thousands of miles away into space. We may be under the impression that it is all safe from sabotage, terrorism, or any warlike conflict. This would be a grave mistake.

Underwater communication cables are secretly mapped very accurately for maintenance purposes and had been held secret for many years. However, such data has been hacked by individuals and sold on. China, Russia and the USA have all designed systems that can knock out satellites thousands of miles up and submersible hardware exist to cut underwater cables. Anti-Satellite systems are known as **ASAT**s for short.

International Space Law (ISL)
ISL is in part an attempt to restrict putting weapons into space. Spying from orbit is allowed, but weapons are not allowed above 100 km, the current legal height where space begins.

ISL is constantly changing as it tries to keep up with technological advances. It includes sections such as ownership of land on the moon; the hardware left there and returned lunar samples.

There are companies that try to sell acres of land on the Moon with full mineral rights. Such offers are actually illegal under ISL. All such offers now have to include the words 'Novelty Item' somewhere on the certificate; although you will need to look carefully with a microscope to see it. But if you paid for a lunar sample return mission from a non-previously landed mission site, then you would own the samples and the vehicle left behind but not that part of the moon.

Dig out a mine on the Moon if you wish and you own everything you find, but do not mine too close to another and cause a collapse. That would be illegal.

What if say China wanted to land very close to the Apollo 11 site, send out a robot to collect a piece of NASA hardware and bring it back for testing. Who still owns it? Would you need permission? Would the footprints left by Neil & Buzz be classed as a preserved historical site? Once messed up by tracks from a rover, that would be it forever. Such points have to be addressed before it actually happens and avoid arguments.

Regarding weapons in space, this so far has been avoided. Satellite killing methods from the ground or instruments deployed below 100 km are allowed but strongly discouraged. Weapon tests have already resulted in tremendous debris fields caused by the destruction of satellites that will litter space for years.

Nuclear warheads in orbit are strictly banned for all nations. This was the greatest fear at the start of the space age as Russia had the capability of launching such weapons while the USA didn't… for a few months.

Earth orbiting objects need to be monitored for the prevention of this very scene from happening.

Air to space tests has included a chemical laser beam fired from a high altitude aircraft to heat up a satellite's fuel tank and explode it. Officially, they were recorded as failures as the atmosphere still absorbed most of the energy before reaching the targets.

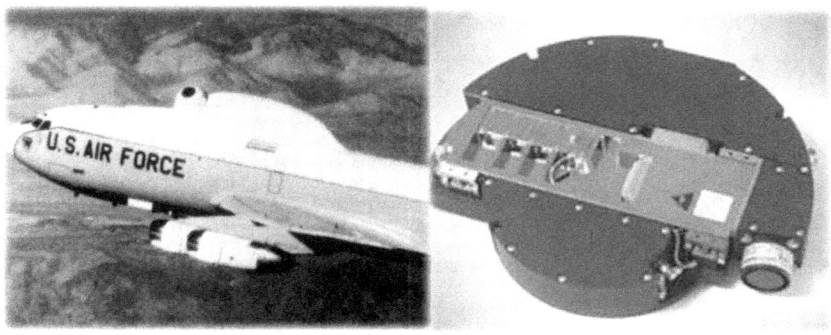

As of November 2014, a new infra-red laser is being tested to operate from high altitude aircraft. The Viper 2.1 was developed by Northrop Grumman and is mainly designed to burn through incoming missiles. If successful, a similar upgraded version should have the capability of destroying satellites.

ISL is a very necessary law standard and many nations have pushed the rules to the limit. It is only a question of time before it is sadly broken.

China's ASAT systems
On 11 January 2007, China successfully destroyed an old Chinese weather satellite. A missile similar in concept to an American system carried this out.

The warhead destroyed the satellite in a head-on collision at an extremely high velocity. Evidence suggests that the same system was also tested in 2005, 2006, 2010 and 2013; although none of those events created any long-lived orbital debris unlike the 2007 test.

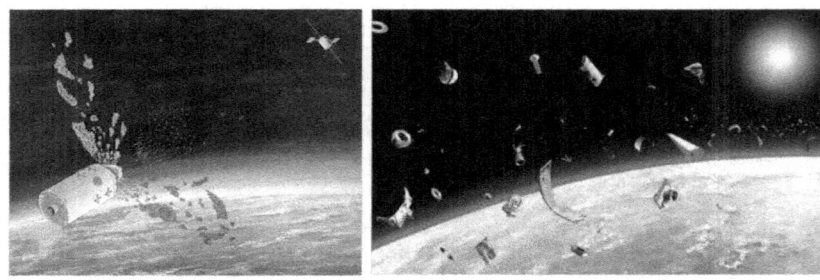

One method of destroying a satellite is to deliberately explode a canister of metallic particles nearby. Such a debris field will hit a satellite and render it useless. The remaining debris will continue on the reach others and remain a threat for decades.

X37B

The USA has a re-usable satellite called the X37B. This is a favourite for Satellite Spotters including myself. It performs in a similar fashion to the space shuttle; it launches as a rocket (well in one), orbits as a satellite and lands as a plane but all without a crew on board. As of 2014, three missions have been carried out with top-secret payloads. The launch and landing always takes place at Vandenberg, California. The X-37B, manufactured by Boeing, weighs five tonnes and measures about 9 meters long, with a wing span of roughly 5 metres.

 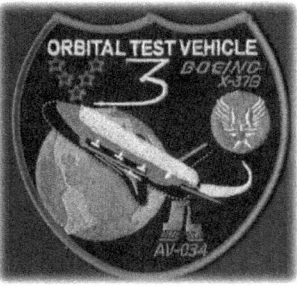

Satellite spotters publish observations on its orbit as it can change without warning. Those with detailed picture taking skills try to discover the payload as it faces the ground. In the past, any satellite spotter who publishes images of such secret missions risked being arrested or at least investigated by the CIA regardless of where they live. All such payloads now are now hidden with a sheet of gold space blanket.

The Pentagon has been asked often if anti-satellite weapons are being developed within the payload bay. They have constantly denied this; other rumours say it may be spying on other spy satellites instead. This sounds more likely.

Within hours, at any time in a high-stake war, the entire Internet and general global communication system can be crippled beyond any rapid repair whether it be cutting undersea cables or destroying satellites in space or both. It would take at least ten years to replace it all. A cyber-attack using software could cause the same meltdown but may only be temporary.

Our Role to make space safe
The International Space Laws are mainly to prevent weapons reaching space. As we can clearly see, year by year, some nations are getting closer to breaking those

rules. Thousands of satellites exist and some can have their orbits changed. Some of them may have the term 'Dead Satellite' officially describing them but may actually be a live one in sleep mode ready for a military test in the future. Others may be described as having a science payload but is a spy satellite instead.

Left; a laser operated by the US Navy. As of December 2014, this is now in operation to destroy drones, aircraft, missiles and small boats. As this can track high-speed moving targets automatically, a similar model could sadly be placed on a satellite (right).

Even NORAD hasn't the resources to keep complete data sets on every target, so we amateurs can play an important role in announcing unusual events observed from own gardens.

www.spacewar.com

Chapter 32   The Future in Orbit

During the early part of the space age, overcrowding in orbit wasn't even considered. The benefits of orbiting instruments monitoring the Earth was not realised some years later. As the cost of rockets and satellites reduced, more were launched.

Today we do face a big problem of overcrowding. This is in various forms; satellites themselves plus flakes of paint, debris from collisions and rockets parts that followed the satellites into orbit. Ideas are desperately needed to reduce this problem. If a critical amount of such material is reached then the entire Earth Orbit field will become chaotic. One large collision can produce 100,000 new pieces travelling at several kilometres a second. These items will in turn destroy other satellites that also break up. This will spread out and continue this chain reaction. This possibility was depicted very well in the movie *'Gravity.'*

Various projects and guidelines have been developed to reduce the impact of overcrowding but remain a major hazard.

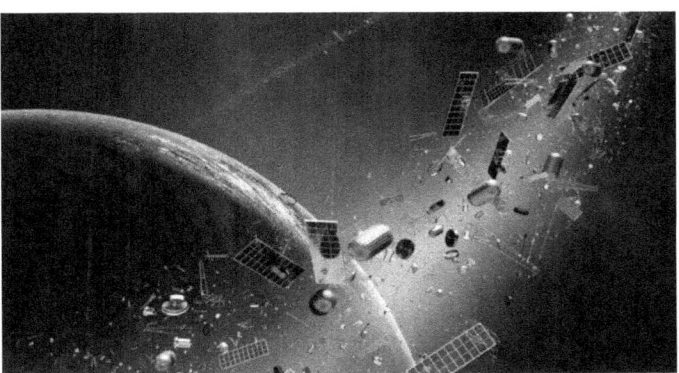

Regulations
All new satellites now have to have a de-orbit agreement. The final commands to it at the end of its useful life is to fire a thruster and reduce its velocity. Within a few hours or days, it will disintegrate in the atmosphere as a spectacular fireball.

Few satellites or upper stage rocket parts that could reach orbit are now painted. In time, paint layers will flake off and become a major hazard. The space shuttle windscreens have been hit a few times by paint and had to be replaced. A slightly harder force would have cracked the glass and breached the vital seal from the vacuum of space.

The Sail Sweep
Engineers have proposed building a sail of several hundred square metres that can be unwrapped in orbit. The sides are kept taught and it simply drifts in orbit; avoiding all the main satellite lanes. Small debris will hit the sail and either vaporise on impact or at least travel through it but loosing much of its momentum on impact. The reduced velocity (Delta V) would force debris into a lower orbit and eventually burn up.

Lasers
Light creates a weak force. A concentrated light source could be projected onto a piece of space debris and slow down its orbit. This would create a low-cost de-orbit system and is currently being tested in Australia.

Longer Lasting Satellites
It has been recently proposed that a robot on a satellite could in theory carry out repairs. Satellites that sit in

geo-stationary orbit are too far out to be serviced by human hands economically. At the end of its normal lifespan these are fired out to a dead zone above and below geo-stationary orbit where they will remain for thousands of years, or de-orbited completely if enough fuel and control remain.

As of December 2014, the US Defence Advanced Research Projects Agency (DARPA) is looking into the concept of placing a robot droid that can move around a satellite and carry out repairs. Upgraded hardware and extra fuel for thrusters could be flown out and the robot replaces the part(s) in orbit. Such droids and robot repair techniques are being tried on the ISS with success.

A droid repairing an antenna on a communication satellite 22,000 miles above the earth. Artist's impression.

Chapter 33 Stamps & Covers

As a super scrimper at heart, there are several other areas that may catch your interest in this field. Stamp collecting is an established hobby worldwide. There are literally millions of special issues produced by most countries. By specialising in one area, a fascinating collection can be built up fairly quickly and cheaply. Just look around at market stalls, boot fairs (meet swaps) museums and hobby stores. The ultimate source I guess by now is eBay. Search for large collections of 'space stamps' instead of purchasing individual stamps or sets. This is far more economical. A collection of 500 space stamps could be purchased for around 1p (1.4cents) each.

They can be displayed in an album in country or by subject order such as Cosmonauts, Satellites, Apollo and Probes etc. Personally, I would choose country as professional investors do.

Covers can be in several forms. These can be released on a rocket launch date for instance, or an anniversary. Many are with no space stamp but issued on the day of an event such as a docking, splashdown, landing etc. A special postmark may be used to cancel the stamps as a mark of the event. Search eBay for 'space cover.'

Paul Weitz flew on the first Skylab mission in 1973. I saw it pass over while he was on board.

Chapter 34 After thoughts on Spaceflight

Many people do often say that space exploration is a complete waste of money, talent and resources. A new hospital can be built for the same cost of one satellite. What they fail to realise is that the one satellite could be studying freak waves and storms in the ocean for instance is saving hundreds of lives. Weather satellites predict hurricane paths and reduce destruction.

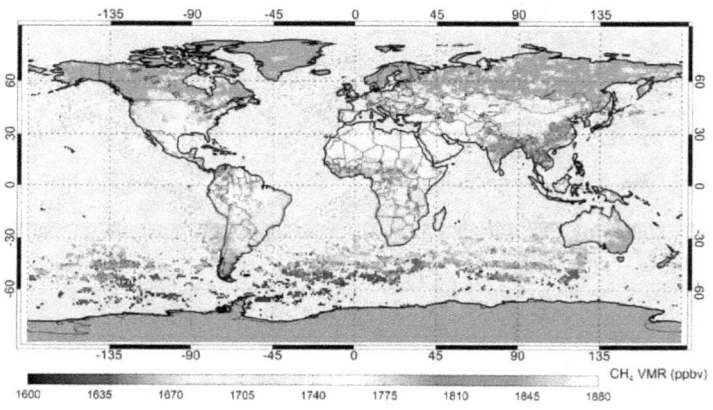

Methane and CO2 emissions are an essential part of climate change. They can only be measured accurately and economically from orbit.

Satellite navigation saves lives on a daily basis also. It has been found to be so crucial, that other systems that are even more accurate are being launched. Galileo is the latest system, ten times more accurate than the US global positioning system.

Satellite technologies reduce the need for hospital beds in the first place; less casualties from storms, aircraft collisions, accidents at sea etc. The cost of construction, launch and operations etc. of such satellites is far lower than the result if they didn't exist.

The Columbus module built at the Kennedy Space Center, Florida in 2007 shot by a typical tourist...me. This was launched to the International Space Station by the space shuttle.

Look around a hospital and note all the high-tech gadgets; most of them were originally designed for spaceflight. For instance, when the J2 engine was built for the Saturn V rocket in 1966, it kept exploding. The engineers knew roughly where the problem existed but as it was so crammed with pipes and valves, the fault could never be found. One engineer thought of building a camera that could produce an image through optic fibres. It was guided through the maze of pipes by a wire and threaded it though. The camera imaged and solved the engine problem on the next test. Today doctors use them and call them Endoscopes.

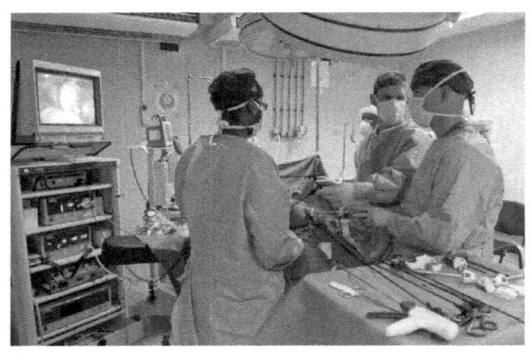

Endoscopes being used here were originally invented for Project Apollo to inspect engines. Other inventions include smoke detectors, WD40, Tempur mattresses and around 100,000 more. Wikimedia Image.

The next image shows rapid ice loss in the remote Arctic and was detected by the Sentinel-1A and CryoSat satellites. Both visible from all over the Earth as they are in polar orbits. Monitoring the Earth from space is detailed, low cost and is safer than trying to gain the same information from the ground.

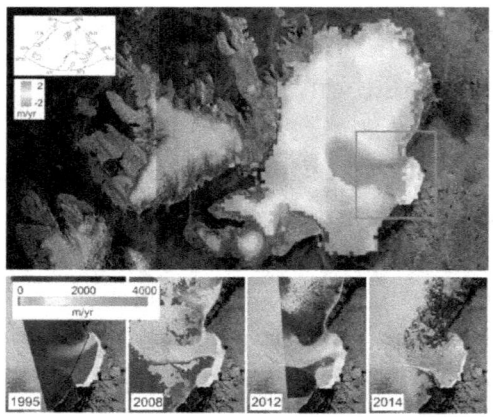

Climate change is affecting farmers; the food grown is becoming more difficult as crop diseases spread.

Satellites can help them by monitoring the health of soil, crops, better use of land and climate prediction.

This harmless and exciting subject can be a real launch pad for some youngsters reading this book. This was the aim from page one.

The world faces huge problems today; Famine, Energy & Material Shortages, Climate Change, Destruction of the Natural World, New Diseases etc; all can be solved by new science and policy. By increasing the number of people involved, we increase our chances of survival or fall upon very hard times and remain stuck here on Planet Earth for centuries. The 21st century is a unique era where we get this choice of futures.

Chapter 35   Further Reading / links

http://www.satobs.org/satintro.html
A starter satellite prediction website.

www.heavens.above.com
The finest satellite prediction website in the world.

www.moonconnection.com
Perfect for Moon phase predictions. During around full Moon only the brightest satellites may be seen.

www.satlist.nl/index.html
A listing of every satellite launched since 1957.

www.satsig.net/sslist.htm
A list of all the Geo-stationary satellites along with the Longitude data. 0 degrees in line with Greenwich, visible from that half of the world. 180 degrees, in line with the centre of the Pacific.

www.tbs-satellite.com
Fabulous site for giving up-to-date technical details on most satellites launched.

www.satview.org/
A real-time tracking website for the most popular satellites such as ISS

www.heavenscape.com
Satellite Tracking Software

www.biotrack.co.uk
Animal Satellite Tracking Company

Book; Observing Earth Satellites by Desmond King Cole 1983.

Book; A vertical Empire – The history of the UK rocket program 1950-1971 by C. N. Hill.

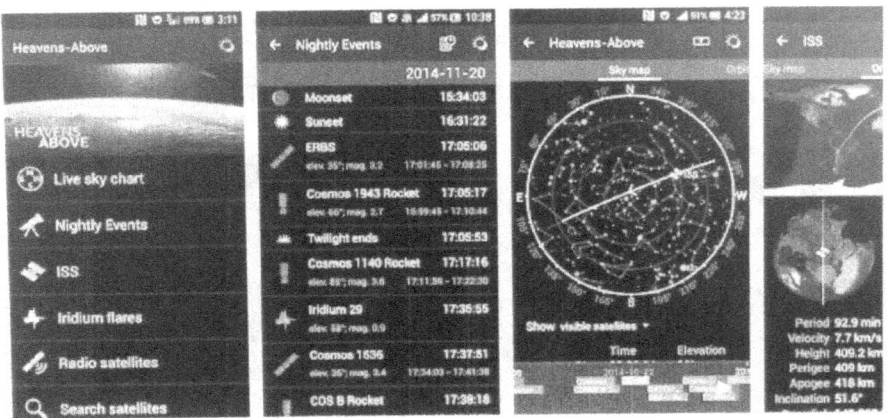

There is a smart phone 'App' that can track some satellites. This is in the early stage of development; they have had reports of inaccurate positioning. The Heavens-Above version has an experimental App for Android users. In time, these will improve no doubt.

Readers; if you have suggestions for future editions of this book please do via

www.outerspacebooks.com

Many thanks… Pete Bassett

Chapter 36   Other Books by the Author

For the very latest listing of all books and versions, use www.outerspacebooks.com

Our YouTube Channel   @astroroadshow

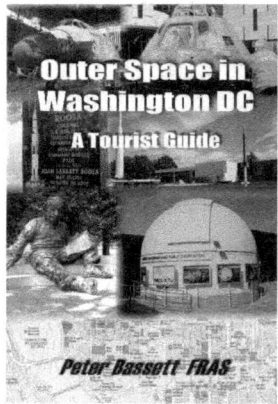
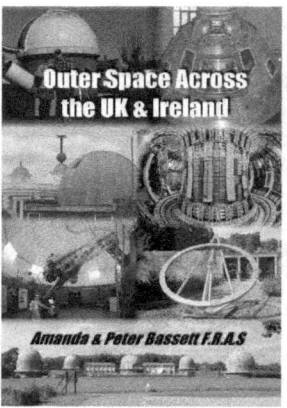

There are B&W, Colour and e-book versions. All book sales support four charities; Cancer Research UK, Kent Air Ambulance, Smile Malawi Orphanage in Africa, British Hedgehog Preservation Society.

www.ingramcontent.com/pod-product-compliance
Lightning Source LLC
Chambersburg PA
CBHW070631220526
45466CB00001B/150